A COW'S LIFE

A COW'S LIFE

The Surprising History of Cattle
and How the Black Angus Came
to Be Home on the Range

M. R. Montgomery

Walker & Company
New York

First published in the United States of America in 2004 by
Walker Publishing Company, Inc.

Published simultaneously in Canada by Fitzhenry and Whiteside,
Markham, Ontario L3R 4T8

For information about permission to reproduce selections from
this book, write to Permissions, Walker & Company,
104 Fifth Avenue, New York, New York 10011.

Art on page 32 is courtesy of Mondadori Electa Spa, Milano.

Library of Congress Cataloging-in-Publication Data

Montgomery, M. R.
A cow's life : the surprising history of cattle and how the
Black Angus came to be home on the range / M. R. Montgomery.
     p.   cm.
   Includes bibliographical references and index.
   ISBN 0-8027-1414-5 (hc : alk. paper)
   1. Aberdeen-Angus cattle—United States—History.
   2. Aberdeen-Angus cattle—Canada—History.   I. Title.

SF199.A14M66   2004
636.2'23—dc22
                                                     2004043073

Book design by M. J. DiMassi

Visit Walker & Company's Web site at www.walkerbooks.com

Printed in the United States of America

2   4   6   8   10   9   7   5   3   1

To Scots and Scotch cows,
indomitable at home and abroad.

"Herdsmen must have a taste and a strong liking to cattle—they must be their hobby."

WILLIAM M'COMBIE OF TILLYFOUR

Contents

Prologue

IN the span of modern human history, a celebrity cow (in the generic sense of a member of the cattle family) is a rare item. Mrs. O'Leary's cow, blamed for the great Chicago, Illinois, fire of October 9, 1871, qualified for a time. Of bulls, only two come to mind, one real, one fictional. They were both famous for being nonfighting Spanish bulls, the real Civilón of Barcelona fame, and for the elementary school set, the marvelously bee-stung Ferdinand the Bull. The Texas longhorn might qualify for celebrity, or notoriety, but only because of western movies and for being the mascot of the University of Texas football team. Both the oaters and the Texas football team are in decline as of this writing. The only other candidate, and it is a winner, is the Aberdeen-Angus breed of cattle.

Across the breadth of North America, only the Aberdeen-Angus have produced a beefsteak that has truly achieved celebrity. Not only is Certified Angus Beef® advertised on the restaurant menu and the supermarket shelf, a picture of an Aberdeen-Angus steer—a silhouette of a solid black hornless animal—adorns menus and package labels in more than five thousand markets and on menus of over fifty-five hundred restaurants.

The simple statistics of this elite piece of meat are boggling. Chosen from just the top 18 percent of all Angus cattle (Americans shortened the hyphenated name for marketing purposes), the certified animals produced in the last year something on the order of 225 million servings of the most desirable cuts of steak, from the gargantuan porterhouse to the relatively effete tenderloin. The rest of these superior carcasses went into a billion and a half servings of labeled roasts and joints and even certified hot dogs. Just to give an idea of how this one particular breed of cow has come to be synonymous with quality, it outsells its only brand-named breed competitor, Certified Hereford Beef, by twenty-five to one.

The success of Angus is not just clever marketing. Smoke and mirrors are of no use with a product that requires only a knife and a fork to easily and rewardingly compare it to a run-of-the-ranch piece of beef. Aberdeen-Angus is remarkable for quality in great quantity and for consistency over centuries. It is one of the few items of food, if not the only one, that can be mass-produced in the twenty-first century and taste as good as it ever did. In an age when gourmets search out tender wild greens and heirloom tomatoes and free-range chickens and artisanal bread and other relics of a less-industrialized era of foodstuffs, there is no need to look for an heirloom steak. Today's Aberdeen-Angus sirloin strip tastes just as good as the famed one eaten by Queen Victoria in 1866. The only difference is that Victoria got hers for free, after having been introduced to the steer by name, before dining upon him.

On the cave wall at Lascaux, France, ancient man painted the animals he hunted, including the now-extinct wild animal that was the source of all of the cattle living today. As varied as our cattle are—from the alert Aberdeen-Angus to the oblivious black-and-white Holstein dairy cow, from huge draft oxen to diminutive Alderney dairy cows—not one remotely resembles that ancestor captured for eternity in the darkness of Lascaux. The bull of the cave paintings survived into the sixteenth century and is the same animal that Julius Caesar described as "a little smaller than an elephant," an animal so different that it is a species separate from the ones we see today. The beast at Lascaux is the largest animal portrayed there, a full 18 feet (5.5 meters) long. The size was an exaggeration, but the artist was conveying the awesome fearfulness of the animal. All the hundreds of breeds of modern cattle have evolved within the last several thousand years, long after Lascaux was abandoned. In Charles Darwin's elegantly phrased book title, they are each a result of *The Variation of Animals and Plants under Domestication.* The horse, it should be noted, unlike the cow, was found in nature of a size and temperament to be ridden or driven and has not changed enough in captivity to be judged a new species.

The cow is arguably the most important animal ever domesticated. It was easily the first, for the bones of cattle, distinguishable by paleontologists from those of its wild ancestor, appear in the garbage heaps of the earliest settlements in the Old World. Cows benefited from and assisted in a revolution in agriculture: people saving seed, planting and harvesting crops, raising animals for meat instead of chasing them through the wilds. When, as one example, cattle reached the British Isles some ten thousand years before the present, the folk, with a more assured food supply, acquired the excess energy to build the great henges—the stone circles—and the barrows and even, in the vicinity of Avebury, to build an entire mountain, apparently for signal fires. British archaeologists once accustomed to thinking of all the ancient monuments as religious sites are beginning to speculate that many of the great circles such as the one at Avebury, with their interior moats and roads, may have also been places to buy and sell cows, "trysts," as the Scots called their gatherings for marketing cattle. The pattern is repeated in many ways in many places, but there is evidence that as a result of the inhabitants raising cattle, a more complex culture was able to emerge, when mankind was nourished and stabilized by the milk and meat of the cow, and assisted in the hard work of farming by the yoked oxen.

Across Europe, hundreds of distinct breeds of cattle developed in isolation, and villagers kept the types they most admired and ate the rest, thus a dizzying array of cattle of all possible sizes and colors and purposes—meat, milk, traction—dotted the landscape. So inseparable were the cattle and the land and the culture of the locals that the breeds came to be called by their native district or hometown: Limousin and Charolais in France; Jersey, Guernsey, and Alderney from the Channel Islands; Herefords from England; and what once were taken as two breeds, Aberdeen and Angus, from those districts of northeastern Scotland. Many breeds of dogs also developed under domestication, but they more often received mundane names from their line of usefulness: pointer (of game birds), shepherd (of sheep), and dachshund (a dog that will go down a badger hole and worry the beast). Geographical dog names tend to be national rather than local—French poodle, German pointer, English setter, Labrador retriever—and very few have the intimate flavor of a cow's prideful town-by-town and district-by-district nomenclature.

It may seem ironic that the Aberdeen-Angus breed, prized for its perceived superiority on the plate, should have originated in Scotland, which is hardly noted for its cuisine, although haggis, made from sheep offal and oatmeal, is actually rather good on a winter night, preferably on Robbie Burns's birthday, and with a single malt whiskey to clear the palate. If ever a country bred up a cow for culinary qualities, France or Italy would seem much likelier candidates. However, in the early nineteenth century when Scottish farmers were refining the Aberdeen-Angus breed, their rivers teemed with salmon and sea trout, the coast bore everything from scallops and cockles to lobsters and langoustines, and offshore dwelt haddock just waiting for a touch of smoke to become the best of all possible breakfast entrées, finnan haddie. Everything was (and in some small Scottish towns still is) all of the finest quality. From the days when their only tableware was a dirk, the Scots have been connoisseurs of great raw materials.

Aberdeen-Angus were not the only cattle raised in Scotland, or even the most common or esteemed. But matters of sapidity and succulence aside, the Aberdeen-Angus cow and her kin have a very remarkable trait—they are adaptable to almost any climate and any style of rearing. For more than fifteen hundred years that are known of, the Scots drove their cattle south to be fattened and sold. In the eighteenth century the Scots began to feed their cattle to fatness and sell them finished, ready for the market. For a number of reasons, to be detailed later, the Scotch farmers kept their cattle indoors during the country's long and wet and dark winters, and no cows could be more content in confinement than the Aberdeens and Anguses. Less than one hundred years later, in a remarkable about-face, the cow that would thrive in confinement became the greatest range cow of all time in the vast North American West, hardy enough to weather the severest cold and wind of winter and the oppressive and humid heat of a High Plains summer.

Through all these changes in lifestyle and migrations to the most boreal and topical climates that European cattle can survive, the Aberdeen-Angus kept their true nature. If handled early and often, they are calm and obedient. If left to themselves, they become self-reliant and, as far as human interactions are concerned, nonsocial. But never antisocial. A free-ranging Aberdeen-Angus wants nothing to do with a visitor, but neither will it viciously attack a human being for no reason. Compared to many breeds, the Aberdeen-Angus retain some qualities one associates with human behaviors and attitudes rather than with those of dumb beasts. Aberdeen-Angus can be variously egotistical, self-actuated, resilient, mothering, sociable (with one another), curious, obstinate, amenable, and even insouciant.

The breed underwent some minor changes in the twentieth century, and most of them, but not all, were improvements. Still, the Aberdeen-Angus has retained all of its great qualities. A time-traveler coming from the Spreyside a thousand years ago would recognize these cattle in the field and on the plate. Its continued success as a domesticated animal seems as assured as anything whose future is reliant upon human beings.

Among the thousands of ranchers and farmers (to use the North American and British words for the same occupation) who perpetuate the Aberdeen-Angus breed around the world, I am proud to be related somewhat distantly to the Felton family of Montana, who raise Aberdeen-Angus. They are not famous breeders and don't want to be. The line between fame and mere notoriety is thin, and the Felton Angus Ranch of Big Timber and Brandenburg, Montana, comes down hard on the side of raising productive, well-behaved (for cows), and self-reliant range cattle—honest, middle-class animals. It is men and women like my cousin Margaret Felton, her husband Raymond, and their son Richard who keep the old Scots values in their herds, a practice fortunately found scattered far and wide across the prairies of the United States and Canada.

✢ I ✢

The Mother of All Cows

UNTIL approximately ten thousand years ago, there was no such thing as a cow. This is not very long ago by evolutionary standards of time, more often measured in units starting at fifty thousand or one hundred thousand years and not infrequently in the millions of years. Within a few thousand years from the birth of the very first cattle, there were cows scattered across the face of the Old World, from Ireland on the west to China on the east. "Scattered" is not a figure of speech. There were and are great gaps in the map of cattle country—inhospitable deserts, Arctic wastes, and tropics with decimating diseases. It is an amazing story, but it is not an evolutionary event in the ordinary sense of that word, for mankind actually invented the cow anew.

Evolutionary biologists agree that all domestic cattle are descended from a common ancestor—one race or another of the aurochs, the primordial ox. That now-extinct beast has the scientific name of *Bos primigenius* (or first cow). All of the humpless cattle in the Western world are of the recent species

Bos taurus. There are slightly different cattle, humped cattle, that may be descended from another kind of aurochs, and those take the name *Bos indicus* because the classic humped cattle come from the Indian subcontinent.

Anyone who has watched a rodeo has seen one of the Indian breeds, the Brahma. They have little to offer in terms of milk or meat (and, of course, they are not eaten at all in India). They were not part of the great story of *Bos taurus,* the prolific genus of all great beef cows and milking cows, until in this century they were crossed with the Aberdeen-Angus. If it wasn't for their unexpected talent for bucking off riders at a rodeo, they would be the least-seen of all cows in North America. Besides a hump and an extremely flexible back that make them all but impossible to ride once they start twisting and bucking simultaneously, all the Indian cattle have long horns and floppy ears. (Most bucking Brahmas have had a foot or two lopped off the ends of their horns to make them less deadly weapons.) Whenever Brahmas are crossed with European breeds, the offspring all have floppy ears . . . it's a dominant gene. When naturally polled or hornless Aberdeen-Angus are bred with any other cow, the crosses are all hornless because the polledness gene is dominant. A registered breed of cows, called Brangus (Brahma × Aberdeen-Angus), are black (another dominant gene), hornless, and floppy-eared. They are a popular, if physically unattractive, beef bred in tropical and subtropical countrysides where the Brahma genes give them some immunity to tropical diseases and the Aberdeen-Angus genes make them palatable.

The world of descriptive biology has scholars who either like lots of species or who like to keep it simple. To the simplifiers (or lumpers) there is no such animal as *Bos indicus*; there is only *Bos taurus* variety indicus, or, in plain English, regular old cows but with humps and beagle ears. The compromisers settle for a subspecies nomenclature and thus, to use a real example, a black Angus cow carries the Latin scientific label of *Bos taurus taurus,* and a Brahma cow works under the alias *Bos taurus indicus.* The purpose of all this Latin is to achieve scientific precision in a universal language. Unfortunately, people do the naming, and nothing human beings are involved with stays simple. Either way, lump or split, there's no reasonable scientific name for the Brangus.

In common English usage, any sort of cattle can be simply termed cow, even if that word has a more precise meaning. Aberdeen-Angus breeders in the West, if they want to be laconic, will say that they "raise black cows." When distinguishing among cattle in a group, a cow is a female that is old enough to have already dropped a calf. A heifer is a young female approaching or just entering breeding age. Males of any age are bulls. Calves are animals anywhere from just-born to sexually mature but not yet involved in breeding. They are either heifer calves or bull calves.

"Ox" was once commonly used in England as a generic noun for cattle taking a singular verb, no gender implied. In North America an ox is exclusively a neutered male draft animal; neutered males headed to the packing plant are referred to as steers. Proper British usage for a steer would be "bullock." "Ox" has now become, in many districts, the British equivalent to our "steer," that is, a neutered male raised for beef.

Heifers are infrequently neutered, and the same verbal adjective is used for them as for cats or dogs; they are spayed heifers. This is a trickier operation than neutering a male and not much practiced. Occasionally, a naturally neutered heifer calf is born because she is the fraternal twin of a bull calf. This is a happening entirely unique to cattle, and the mechanism is not understood. Apparently the male hormones circulating in the mutually shared blood system of the cow and the calves effectively sterilizes the female, whereas the male is always normal. These sterile females are called freemartins, a term whose origin is lost and which has no other meaning. Freemartins are highly prized beef animals, tender and well marbled.

.   .   .

There is no record, no physical evidence, of how and precisely when and where Stone Age humans tamed and bred the aurochs. What is certain is that they did it very carefully because the aurochs was an animal nearly twice the size and many times the weight of the hugest modern cow. Worse, from Neolithic man's point of view, the aurochs had a reputation for fierceness equal to that of bears or lions. There are very few contemporary accounts of aurochs, and none of them encouraged getting close to the beast.

The once-well-known description of the aurochs is in Julius Caesar's *Gallic Wars,* where he mentions several unusual animals found only in the primordial forests of northern Europe and Britain. The aurochs was just such a curiosity; the animal was long since extinct anywhere south or east of modern Germany. It is also the only strange animal Caesar wrote about that actually existed, for Caesar was as gullible as the average tourist in the American West who believes in Jackalopes. This is Caesar's account of the aurochs:

> *In size these are somewhat smaller than elephants; in appearance, color, and shape they are as bulls. Great is their strength and great their speed, and they spare neither man nor beast once sighted. These the Germans slay zealously, by taking them in pits; by such work the young men harden themselves and by this kind of hunting train themselves, and*

those who have slain most of them bring the horns with them to a public place for a testimony thereof, and win great renown. But even if they are caught very young, the animals cannot be tamed or accustomed to human beings. In bulk, shape, and appearance their horns are very different from the horns of our own oxen. The natives collect them zealously and encase the edges with silver, and then at their grandest banquets use them as drinking cups.

Temperament aside (and exaggerating the ferocity of wild animals is a universal and continual human habit), the sheer size of the aboriginal aurochs militated against domestication. There are no examples of large, wild animals successfully domesticated—not even the "tame" Indian elephants used for logging in Asia. Those elephants were captured wild and young (being unwilling or unable to breed in captivity), and training them is one of the most dangerous of occupations. Circus elephants and such obviously dangerous animals as circus lions and tigers are not domesticated, but they have been behaviorally modified one by one. The offspring of the "tamest" tiger (or elephant), left to its own devices, will be as deadly as a perfectly wild animal. Dangerous performing animals require a regular and continual regimen of training and reinforcement.

To this day, all of the great animals of Africa—from zebras to elephants, wildebeest to elands, dogs and wild cats and dozens more—remain undomesticated. Only a handful of even moderately large animals have been domesticated. The wolf and probably some other wild canines became our dogs, and some dogs are the tamest, most domesticated of all pets. Other large wild animals that have been bred and selected through centuries include the European wild boar, which became our pig; a single species each of wild goat and sheep; the camel (and its American cousin, the llama); and, of course, the cow and the horse, the largest European and Asian domesticates.

The first horses ever tamed, almost certainly in central Asia, did not appear to have to be changed at all; since they ran wild, they were tamable, and once tame, they became a domesticated animal without enough physical change to make them a new species. The domestication simply repeated itself in the New World. North American native peoples, who had never seen a horse before, managed to capture and tame feral Spanish horses that had long since escaped from the conquerors of Mexico and had multiplied and spread into the northern plains, hundreds of miles from the nearest European settlement.

It was no lack of intelligence or culture in Indian country that kept emigrants from seeing dairy herds of tame buffalo on the way west. The fault is in the animals. That is the difference between an animal's ability to be domesticated or to remain perpetually wild. The capacity for tameness has to be bred in the bone. Mankind created the modern cow, goat, sheep, and pig (a sometimes quite dangerous domesticate), but we have never gotten past efficient killing when it comes to managing a bison.

Some of the reasons for this lack of domestication in African and North American fauna, and its consequences, are elegantly summarized in Jared Diamond's *Guns, Germs and Steel: The Fates of Human Societies.* The bottom line is apparently this: A capacity for calmness is the first requirement in a domesticated species. Wolves are both calm and playful except while eating an elk, and even then, they share. Skunks, on the other hand, although slow-moving, are nervous. People do keep skunks in their houses, but they are impossible to litter-box train, and they won't come when you call them. These "pet" skunks (usually with their squirt glands removed) are neither tame nor domesticated; they are commensal—they just happen to live in a house.

Some cats and most dogs are symbiotic; either man or beast gains something from the relationship.

Another critical element in domestication is that the animal must have some kind of group culture in the wild; it must naturally form larger groups than pairs or families. In short, they must run in herds or extended family groups that have a confirmed social order. (A lack of any group social order is another reason that cats are so problematically and barely domesticated.) The vast herds of African animals, wildebeest and zebras, for example, are aggregations of individuals that, although they move in groups, avoid contact and have no social structure governing them.

Whether ancient people solved the problem of behavior first or reduced the aurochs to a physically manageable size and then gentled it is, of course, unknowable. Certainly reducing the enormous size of the ancestral animal would be one priority, for its mere bulk and strength would militate against a breeding program. The cow and the horse are at the absolute limit of the size of animals whose breeding has ever been controlled by human beings—outside of a zoological park. So, logically, a first step was to get the animal down to manageable size, and it was a long, long way down.

To get some idea of what an aurochs looked like, how much it weighed, its height and length, paleontologists have to work from an existing, living model. Based on the very lifelike cave paintings of aurochs (and the testimony of Julius Caesar), it is assumed they had the general proportions of our fighting bull, particularly the large head, bulky withers, and long dorsal spiny processes that support the head and horns. The closest living creature in the shape of an aurochs is the Spanish fighting bull. While other breeds have been created for milk or meat or locomotive power, the fighting bull has been built for speed, endurance, and, most particularly, for attitude. Spanish authors of books on cows even create a special subspecies (although without any scientific underpinning) and try to pass the fighting bull off as *Bos primigenius iberica*, the Spanish aurochs. The fighting bull, *toro bravo*, is clearly a recent development springing from ordinary domestic cattle, a fact reflected in its variable colors. Very few wild animals are different-colored from their brethren. Predatory animals seem to have more flexibility: Bears, wolves, foxes, and panthers all come in two or three colors, from black through brown to almost blond or silver. Although fighting bulls are commonly black, that is a modern affectation. The original bullfights were against more colorful and variable animals.

The fighting bull's horns differ so much from animal to animal that Spanish has more than a dozen names for the various curvatures and lengths of the horns. Such variations do not occur in wild species; there, odd-shaped horns are the idiosyncratic result of a genetic fault, injury, or aging. The occasional Spanish bull that does have horns somewhat like an aurochs's is called a *corniabierto*, literally "open horned." Aurochs's horns curved gracefully forward in a large arc, the way a human being's arms would curve around a 55-gallon (140-liter) barrel. The aurochs carried its horns almost parallel to the ground (and its own jawline), more like an African cape buffalo than a cow.

Despite similarities of physique, the aurochs was much larger than the Spanish animal. The largest bull ever to enter a ring was Cocinero; why the bull was nicknamed "cook" has passed from human memory. The term may have been a comment on the tendency of cooks to gain weight. Cocinero fought and died at Malaga on June 3, 1877. According to the rules of bullfighting, he was weighed in the bull ring's shambles after being gutted, skinned, beheaded, and having his hooves cut off.

Dead and ready for the butcher shop, Cocinero weighed 414 kilograms, or 912 pounds. Since the dressed weight of a fighting bull averages about 65 percent of its live weight, Cocinero would have weighed 636 kilograms, or 1,402 pounds, alive. Beef cattle can easily weigh that much—that would be at the high end for a feedlot steer going to slaughter—but they are chubby, whereas the *toro* is leaner and meaner. Fighting bulls run in height, measured at the shoulder—but without taking into account the swollen neck and withers of an enraged animal—less than a meter, say, 32 to 34 inches. They give the impression of greater height, but closely inspected photographs of a bullfighter passing the animal next to his body indicates their true height. The bull usually comes up to the *torero*'s waist, no higher. The commonest place to be gored is from the knee to the belly button.

It is possible to gain a good estimate of the height and weight of an aurochs by scaling up the bones of the fighting bull to match the fossilized bones of the primordial ox. The usual com-

**Size comparison of a modern bull,
an aurochs, and a bullfighter**

parison is made of the large metatarsal bone, the bone between the hock and the hoof. The results are astonishing.

A basic principle involved in weighing an imagined (but once very real) aurochs involves the rule of calculating volume in three-dimensional objects. The surface area of an animal (think of its skin stretched out flat for easy measurement) increases by the square of the dimensions (a hide or a rug two feet square equals four square feet; a four-foot-square rug or hide totals 16 square feet, etc.). A measurable change in the height or length or an animal will increase its surface area by the square of the proportional increase. But volume is another matter; volume changes by the *cube* of the dimensions. A one-foot-square block of ice is, of course, one cubic foot. A two-foot block has 8 cubic feet and a three-foot block has 27.

The same principle applies to very complicated shapes, like cows, if the proportions among all the parts remain constant. And since animals have a specific gravity approximately equal to one (1) no matter what their size, the cow scales up in weight almost as neatly as a geometric solid like a block of ice. That makes it possible to re-create, and then "weigh," an aurochs. Paleontologists do this by considering the relative length (or width or circumference, it doesn't matter) of bones that are anatomically identical in *toro* and aurochs.

To estimate the overall dimensions of an aurochs, we need to assign an average height and weight to the fighting bull. For convenience, and to err on the side of underestimating the size of an aurochs, a fighting bull will be defined as one meter (39 inches) tall, which is very tall for the breed. This will actually *underestimate* the size of an aurochs because it will decrease the proportional, scalable differences. The British aurochs is one and two-thirds larger in height than the one-meter fighting bull; the continental ones are from one and three-quarters to twice the height. But we are also interested in the sheer volume (and the weight) of our aurochs, and for that, we have to remember to cube the difference in the dimensions. Fighting bulls vary, but 500 kilograms (1,100 pounds) is a convenient and reasonable number. The legal minimum to enter the ring in Spain is 470 kilograms, or 1,036 pounds, on the hoof. To estimate the weight or mass of an aurochs, multiply the modern animal's weight by the difference in dimensional proportions *cubed*.

Fossils reveal that the aurochs' size varied a little from region to region and by conflicting interpretations of scientists. English aurochs seem to have been a little smaller than continental animals. The average English male beast was some 1.6 meters (5 feet 3 inches) tall. Various European aurochs have been estimated (again, for the male beast) at 1.75 meters (5 feet 9 inches) to as large as 2.0 meters (6 feet 6 inches).

The British aurochs, smallest of the ancient ones, would have weighed about four times (1.6 cubed, or 1.6 × 1.6 × 1.6 = 4.096) as much as a *toro*, or 2,048 kilograms, 4,514 pounds, a whopping two and one quarter tons of mobile and hostile beef. The continental aurochs, by the same formula (dimensional proportion cubed), weighed from 2,680 kilograms to 4,000 kilograms (from almost 6,000 pounds to some 8,000 pounds). It appears that Julius Caesar was not exaggerating when he said they were just a little smaller than an elephant.

·   ·   ·

Using bone size to estimate a whole animal's size and particularly its weight may seem a little theoretical, but Aberdeen Angus breeders know it is practical. Every spring and fall, all across North America, some breeders will do precisely what the paleontologists do, measure a bone and define a weight based on that bone. One of the many tedious tasks in raising registered beef breeds is the obligation to keep a record of birth weight for each animal and the beast's weight at weaning. (The second weighing shows rate of growth.) These are important numbers because the perfect beef animal for breeding turns out calves with a fairly low birth weight (this makes for easier calving) and a high wean weight (which demonstrates a genetic tendency for quick growth). Newborn calves, less than two days old, are reasonably easy to catch and hold for tasks like putting a numbered tag in their ear. (After a day or two, some serious foot chases take place.) Tagging goes smoothly most of the time; the calf is pinched between the rancher's legs for the few seconds it takes to pin the tag through the ear. Weighing is harder. The calf has to be placed in a sling or have two legs tied together so that a scale can be hooked onto the critter. Hoisting it clear of the ground is not fun: The rascals weigh anywhere from 75 to 110 pounds (34 to 50 kilograms), and their longish legs make it necessary to hold the scales almost up at eye level to get them off the ground.

Researchers in Ames, Iowa, developed a simple cloth tape measure for newborn Angus calves—one side for bull calves, the other for heifers. The rancher can use it to measure the outside diameter of a rear leg "ankle" (just above the hoof where the metacarpal and the proximal seismoid bones overlap), and the tape measure gives the weight in pounds. The Ames researchers guarantee getting the weight within three pounds every time. This is the same conversion technique paleobiologists would use to calculate the mass of an extinct animal.

What applies to oxen also applies to scaling up human beings. The chart in the doctor's office suggests appropriate weights for a person of a specific height. These weights, somewhat specious, are calculated by cubing the increase in size for each increment. And the rules hold well even in extreme cases. Robert Wadlow, a well-known British giant of the pre–World War II era, measured 6 feet in height (1.8 meters) and weighed

169 pounds (76 kilograms) as a twelve-year-old. Shortly before his death at twenty-one years from an accidental fall, Wadlow had reached 8 feet 6 inches (2.6 meters). His height had increased by a ratio of 1.4 to 1 since his measurement at age twelve. The ratio cubed is 2.98. If scaling worked perfectly, his weight at death should have been 2.98 × 169 pounds (or 76 kilograms) = 504 pounds (228 kilograms). In fact, Wadlow weighed 491 pounds, or 222 kilograms. The estimate was off by a little less than 3 percent.

Given the huge mass of an aurochs, holding the animal in an enclosure would have required very cleverly designed, strong fences, even if all the animal did was lean against it. Sheer weight, of course, does not necessarily make an aurochs more dangerous, but mass in motion does. The aurochs' increased height meant increased leg length and a longer step compared to a modern bovine. Animals of the same body type tend to gallop at the same rate (strides per minute), but the distance covered in that minute by each stride is what makes the *speed*. That is why ostriches run faster than pheasants, and horses always beat ponies to the finish line. The speed of an aurochs is truly difficult to estimate, but it is surely greater than that of a fighting bull, a bison, or an African Cape buffalo. Bison can run up to 35 miles an hour (56 kmh). Two or three tons of aurochs would have come hurtling at Neolithic man (or a fence) at some 45 miles per hour (72 kmh), courtesy of its longer legs. The speed matters even more than the weight. Being hit by a freight train going 1 mile per hour (1.6 kmh) might leave one slightly bruised at the point of contact, but a skinny messenger speeding along on a bicycle can kill someone on impact. It's the combination of weight and speed—the kinetic energy, calculated at one-half of the weight times the square of the speed—that kills, not the mass by itself or the motion alone.

Typically, paleontological calculations are made using

heights, since a single leg bone is all that is needed, not a whole skeleton. But for very rotund and short-legged animals, like the hippopotamus, differences in total body length (head plus body) work the same magic as height in other mammals. With extinct animals, a researcher would use the length of a vertebrae or two and a skull to estimate the total length. The pygmy hippo averages about 5 feet 9 inches (1.75 meters) in length and weighs some 700 pounds (270 kilograms). The full-size hippo averages 15 feet (4.6 meters) in length, or 2.6 times its pygmy look-alike. The average regular hippo comes in at 8,800 pounds (4,000 kilograms). Using the formula of cubed proportionality, you would get a predicted weight based on the pygmy of 8,500 pounds (3,860 kilograms) for the standard hippo.

The massive size of the aurochs adds to the mystery of how our ancestors, sometime in the Neolithic Era, managed to scale down the aurochs into domestic cattle that were much smaller than most of today's breeds and induce them to put up with close contact with humans. The usual explanation is that somehow we managed to selectively breed them to the desired size, just as we made wolf-dogs smaller.

One theory is that after the last ice age, the aurochs naturally shrank, and humans domesticated the diminished animals. The scattered bones of aurochs suggest that the ones found in the Paleolithic (old Stone Age) assemblages of human garbage were some 10 percent larger than aurochs from early Neolithic (new Stone Age) trash heaps. In other words, over some 600,000 to 750,000 years from the earliest age of stone tools to some 10,000 years ago, the aurochs shrank in size by 10 percent.

In the next two or three thousand years, by the Late ("Pottery") Neolithic, cattle (*Bos taurus*) suddenly appeared. They were one-third to one-half as tall as the ancestral aurochs, had small horns, and, more important in genetic terms, the horns were *variable* horns. They had become a new species. This occurred over a very short period of time. Of course, this change didn't solve the problem of how to get started working with a three- to four-ton beast that is quick-footed and furious.

To selectively breed animals of any kind, one has to have control over them and particularly over their sex lives. How this was done with such a huge beast is mysterious.

According to their earliest fossil bones, cows were comparatively diminutive and easily distinguishable from aurochs. Remarkably, there are no known intermediates between the big aurochs, the slightly smaller ones later, and the modern cow. Such is true with many evolutionary series but uncommon with a relatively recent event. Still, most paleobiologists continue to theorize that with the end of the last ice age, the aurochs became much smaller without human intervention, as a result of no longer needing great bulk for temperature regulation. (Northern animals tend to be larger than similar species from temperate zones; and often within species, say white-tailed deer, animals from along the U.S.–Canada border tend to be much larger than their counterparts in Louisiana or Mississippi.) But this is hardly convincing: The Aurochs' most similar contemporary, the European bison, or Wysent, didn't shrink an inch when the last ice age ended.

One intriguing possibility is that humankind occasionally encountered miniature aurochs. There are two documented events in the history of the genus *Bos* to look at for evidence. First, in Central Europe, aurochs' skulls have been found that are "about one-third below the normal size, but do not differ from ordinary skulls in other respects." Scientists under the sway of old-fashioned, slow-but-steady evolution dismissed these mini-aurochs as aberrations, perhaps resulting from malnutrition, although there is little evidence anywhere in biology that malnutrition diminishes an organism by 33 percent. The worst sort of nonfatal human malnutrition might produce an adult between 4.5 and 5 feet tall (1.37 to 1.52 meters), a diminution of less than one-fifth the size of normal human beings. This is hardly the reduction in size seen in the mini-aurochs fossils. Furthermore, a starved animal—human or otherwise—would not have the normal bones found in the diminished aurochs. If a smaller animal was born in captivity, it might well survive better on its restricted diet. This would be a form of (artificially induced) natural selection. But it is the "born in captivity" that is so problematic, given the size of the wild animal. In places in the world where tribes are isolated and food is scarce, some naturally selected pygmy humans have evolved. But by no means do they approach the relative smallness of the post–Stone Age cows compared with aurochs. The famous African pygmies range from 4 to 5 feet tall (1.2 to 1.52 meters), a mere decrease of at most 20 percent compared with other humans. That doesn't approach the more than 50 percent decline in height from aurochs to Aberdeen-Angus.

What may have happened to some aurochs, and what has happened in modern time within the genus *Bos,* is the sudden appearance of miniature calves as small as one-third the size of the parents. These tiny cattle come in two distinctly different forms. In one case, there are dwarfed calves, usually unhealthy, malformed, and prone to early deaths. This kind of misshapen dwarfism is controlled by a recessive gene. A second type of dwarfism produces more symmetrical midget calves. The best-documented case was of a Charolais bull (no name but numbered 717705) at the United States Meat Research Center in Florida. He produced numerous miniature, well-formed calves, regardless of the kind of cow he was bred to. The only explanation was that No. 717705 had a defective (or dissimilar) gene among the many genes that control the size of animals. And it was a dominant gene. Every cow he bred successfully turned out a midget. If No. 717705 had not been destroyed, he would have continued to sire calves eerily reminiscent of the first cows compared with aurochs—half or less the size of the existing breeding standard. More important, his offspring, particularly if bred incestuously, would have continued the miniature line for ages and ages.

In Canada a dominion research station actually collected miniature cattle, Herefords for the most part, and started a small herd of very small cows in the early 1950s. When they realized there was no point in making pint-size rib roasts, they terminated the experiment. To this day, a visitor to a big animal exposition may encounter a pen of midget Aberdeen-Angus, little one-third- to one-half-size replicas. The one thing the visitor will never find out is their ancestry. As with most freak show attractions, mysterious origins are part of the act. They would certainly have come from a "commercial" herd, not from a registered herd, because the sources of dwarfism are too well known, and the registry system too exact, to allow a pen full of miniatures to develop by accident in a registered herd. The last Angus bull known to carry a gene for dwarfism was calved in 1977, and none of his offspring, male or female, can be found in a registered herd today.

Given this plasticity by mutation in the genus *Bos*, it is possible that a Neolithic person, wandering through the forests and glades of his homeland, might have happened upon a very small aurochs calf and brought it back to camp. This would be an excellent start on the creation of *Bos taurus*, but not the entire answer. There are any number of animals no larger than a domestic cow—bison, Cape buffalo, zebras, wildebeest—that have never been brought under any semblance of control. What was also needed, and perhaps this began while the aurochs was still enormous, was to create a docile beast.

.   .   .

Just as the genus *Bos* has shown extreme variability in size, including spontaneous dwarfing, it has also produced individual animals with temperaments quite different from the parent stock. For an inexplicable sudden change from violent behavior to extraordinary sociability, the best-known example was the famous Spanish fighting bull Civilón, who lived in Spain just before the Spanish Civil War. He behaved so tamely that small children flocked to the fighting bull ranch to approach this marvel of nature and offer him handfuls of succulent grass and edible bouquets of wildflowers. His sobriquet translates approximately as "Big Civilian," for he lacked the "military" ferocity of his herd-mates. Famous across Spain, where his photograph with attendant children was a staple "human interest" newspaper story, Civilón was finally consigned to the bull ring in Barcelona, as a rather morbid spectacle. The city was threatened with attack by Franco's fascist forces, and siege conditions are bad for the human psyche. A promoter thought the spectacle of a pacifist bull would intrigue the public and distract people from their problems.

A full house came to see if the mild-mannered beast would fight. Fight he did, when the mounted picadors lanced him between his shoulders to get him to drop his head so that, when the moment came, the *torero* could drive the sword home, leaning over the bull's horns. He was as brave as bulls can be, repeatedly charging the picadors, ignoring the pain, toppling horses one after another, chasing the picadors behind the barricades. He was granted, as rarely happens, a reprieve from facing the matador and the sword. Civilón had earned an indulgence, *un indulto,* for his bravery. His keeper from the ranch walked out onto the sunlit arena, whistled him over, and walked Civilón out of the ring and into the stables, where he could recover from the *pics* before being shipped back to the ranch and a well-earned retirement. Unfortunately, on the night of July 18, 1936, Franco's Falangist rebels entered Barcelona, and looking for food, ransacked the bull-ring stables and ate Civilón for breakfast. Although the rebels were driven out of the city on July 19, it was too late. The date when they ate the gentle beast is the one generally regarded as the beginning of the full-scale Spanish Civil War.

. . .

While an occasional gentle aurochs would have helped the process of domestication (and transformation into a new species), the whole process would have been much more rapid if a small group of aurochs simultaneously became less fierce. This moderation of behavior in large unfriendly wild animals, called for lack of a better phrase self-taming, does occur. It has been studied most extensively with North American wild sheep and suggests that there is a distinct possibility that the aurochs met humans halfway or more in a kind of auto-domestication. Another example has been observed in Africa by a biologist acquaintance of the author with much African experience. He related that in two game parks (large outdoor zoological garden/wildlife refuges) some self-taming behavior has been noted with the dreaded and ostensibly fierce Cape buffalo. He personally observed Cape buffalo that had gotten so amicable in one game park that two wardens could sit facing each other on a bull's back and play gin rummy on the impromptu card table.

The ethologist and observer of wild animals, Valerius Geist, related a remarkable example of the ease with which aggressive and territorial mammals may be tamed. In his *Mountain Sheep and Man in the Northern Wilds*, Geist recounts discovering not one but two bands of bighorn sheep in British Columbia that had been, before he arrived and unknown to him, tamed by local game wardens. In both cases, the wardens had used salt as the lure and the reward for good behavior. Sheep, like all browsing animals in cold climates, lose bone minerals during the winter as they barely survive on a scanty diet. The associated minerals found in unrefined salt (sodium chloride plus traces of several other metallic salts) and the minerals in salts blended specifically for animals are naturally desirable, even essential for life.

Imagine, as Geist found himself, standing on a precipice where a fall meant certain death and being charged by a band of bighorns led by two massive rams. And imagine them skidding to a stop and waiting patiently for this new two-legged salt-provider to be nice. Before Geist discovered who had tamed the animals and how it had been done, he found himself in constant social contact with the wild beasts. When he began, on a hunch, feeding them salt, they fairly fell in love with him. The biggest rams would lie down next to him, actually touching him as he sat quietly. Geist took great amusement from leading a neophyte biologist out onto the mountain, telling the nimrod how dangerous the bighorns could be when defending their territory, and then watching as a ram known to him by name charged over to bum some salt.

I informed them of the terribly dangerous ram that roamed the mountains who took out his vengeance on people on sight. Each time Crooked Horn did me the favor of appearing, and each time, as soon as he spotted us, he came at a run from about two hundred yards away. A big ram approaching at full blast is a sight that causes [an] anxious moment in the

*hearts of even brave men. When Crooked Horn was about
150 yards away, I yelled "It's him! It's him!" What splendid
results! My pals took to their heels, lickety-split, down the
mountain. What a sight it was to see a dedicated hunter running as hard as possible from a fine bighorn ram.*

If aurochs were held, as some paleontologists speculate, for
ceremonial sacrifices in Neolithic villages, perhaps a few animals gentled themselves, owing to the provision of a little salt
or something else the beast craved. But working the magic of
reducing even tamed aurochs to less than half their wild dimensions required gaining control over breeding. Neolithic
man would have had to select the smaller, easier-to-control aurochs and mate them deliberately. That implies fences that
would not only keep swift two-ton animals inside but also keep
the wild aurochs bulls outside the corral. It is possible that our
late Stone Age ancestors could have achieved all of the construction skills and practical understanding necessary for selective breeding. After all, some ten to twelve thousand years
ago they would take up agriculture and build watercraft, each
a great leap forward. But it seems unlikely that they would be
able to impose their will without the concurrence of two spontaneous events: genetic changes that created smaller animals
and some kind of auto-taming and self-domestication within
that same population of wild aurochs.

The full impact of human activity on the evolution of the
aurochs into modern cattle may never be known, but there is
less uncertainty about the effect of the cow on the human condition. It is no exaggeration to say that cattle have changed human beings, not physically as we changed them but culturally.
Long before there were Highland Scots enamored of their Aberdeen and Angus cows, the cow had to change the world so
that there could be Scots.

⇥ 2 ⇤

How Cows Changed
the World

MOST of the world's advanced cultures, ones that eventually had written languages, developed in the company of a multitude of cattle—cows and bulls of many sizes and many colors. Put another way, except above the Arctic Circle or in the most disease-ridden tropical lands, anyone reading this book in a bookstore or a public library will be within a half-day's drive (often much less) of a cow. The history of what we think of as civilization is, with very few exceptions, a story intertwined with cattle, a narrative pulled along by oxen, a growth nurtured with butter and cheese. The first known towns established in Europe, North Africa, the Middle East, and the Far East—that is, life outside of a cave—were places with cows. Successive waves of migrators (or invaders) across Europe and Asia brought cattle with them.

It is not just in Western societies that cows mattered. The very earliest pictographs from China and Egypt include ones for cows and calves or bulls. There are paintings of milk cows on Egyptian tomb walls and bas-reliefs of bulls on the arches of Babylon. All these places and more had cows before written language, before cities, before the wealth to become conquerors. The first rudimentary symbolic or pictorial written languages had an image that stood for cattle. The man on horseback might be the warrior, but the capital to own a horse and equip an army depended on the cows at home. The Greeks who laid siege to Troy are best remembered for chariots and horses and, of course, the Trojan Horse. But they sacrificed, and feasted on, cattle.

Cows were the first large animals that we domesticated, our first traction-power for plowing. Rudiments of agriculture preceded the plow, when people first scratched at the soil and planted seeds. Add cows and plows, and a revolution followed. The milked cow was our first renewable and reliable source of protein and fat. Cattle accompanied the spread of farming across Neolithic Europe and Asia. Not a year goes by but archaeologists gain new insights into the importance of cattle in the ancient world. Recent research by scholars at the Harvard Semitic Museum indicates that the builders of the Egyptian pyramids were not slaves, in the sense of imprisoned and mistreated workers, but young, well-fed Egyptians who had a diet rich in beef. The area around the pyramids holds layers upon layers of cattle bones, buried under the sands. At the western edge of Europe, British scientists have been able to find indisputable chemical evidence that the oldest pottery shards in England, from around 6000 B.C., once held milk, likeliest in the form of butter and cheese, two ways of extending the "shelf life" of milk.

Wherever our ancestors raised cows, they created or adopted a plethora of alphabets and symbolic characters, mathematics, and literature. Cultures without cows, as in the Americas and sub-Saharan Africa, relied on conquest and slavery to create wealth. In Old World landscapes unsuitable for cows, camels, goats, and sheep sufficed to get a civilization under way but not to fatten it up properly. The great era of Arabic intellectual culture thrived in places with cows—Alexandria in Egypt, Istanbul, and the Moorish cities of Spain. In the remote and inhospitable mountains of Tibet a cousin of cows—the yak—against all environmental odds supported a society capable of building astonishing Buddhist temples and inscribing prodigious numbers of religious documents.

In all of Europe and most of Asia and North Africa, the cultures with cows replaced those that had none, driving them to the very fringes of habitable land and overwhelming them completely. Much is made by anthropologists of the "agricultural expansion" that radiated out of the Middle East, creating advanced cultures from China to Ireland. And just what radiated? Grain crops, to be sure, and just as important, cattle. All of Gaul (of Celtic Europe) might have been divided into three parts by Julius Caesar, but long before he got there, all of Gaul was cattle country. And when Caesar invaded Britain, Bodacia, warrior queen, tried to push him back into the sea. The legend says that she led her army in person, driving a chariot pulled by battle-cows.

We talk about the great inventions that mark the advance of human cultures . . . the Bronze Age that replaced the Stone Age with deadlier swords and better tools, the Iron Age that gave us the weapons that sliced through bronze shields and the durable tools to manufacture ever more complex objects. The era of iron became so complex after many centuries that it segued into the Machine Age, the very modern world as we know it. After World War II came the Atomic Age. But before all this weaponry and manufacturing technology, the first period of overwhelming change, of movement toward a host of literate and mathematical civilizations in the Old World, was the Cow Age.

Reading a flat statement that advanced cultures depended on the domestic cow, a skeptic might think: "What about Genghis Khan? He had nothing but horses when he conquered China, and then his heir Kubla Khan came up just a typhoon short of capturing Japan, didn't he?" That is true, but it was a very brief era, and even within those two hundred years, that Mongol horde started to become Chinese. The culture with cows started to rub off on the foreign conquerers, and the Mongols in China lost the respect and the alliance of the other Mongol hordes. "Too Chinese, too civilized"—that was the judgment of their peers. Another horde pillaged Europe, all the way to the gates of Krakow, and none of them stayed. They were just a plague, a wildfire of destruction that swept all before them and left nothing behind except the dead.

Westerners are unlikely to think of the Far East as "cattle country," but the value of the cow is indelibly inscribed in both Chinese and Japanese thought. The earliest form of written Chinese, established centuries before Genghis Khan's invasion, had a pictograph for "cow," which is pretty obvious, right down to the lyrelike horns so typical of ancient cattle, as depicted on everything from Egyptian tomb walls to Greek vases. It is unlikely that the world was occupied entirely by cows with recurved horns, a rarity today. The curious horns seem to be the result of the great difficulty primitive artists had with perspective. The horns of aurochs drawn on cave walls are also lyre-shaped, but the fossil remains of their horns have simple curves.

Written Chinese evolved from obvious pictographs to subtler ideographs. An ideograph is more than a simple representation: Its fuller meaning can include some of the attributes of the object—in this case, the serene power of a cow. Somewhere along the way, the cow lost one of its horns but none of its symbolic power. The character for "cow" becomes a superlative when you put it together with another ideograph. In Chi-

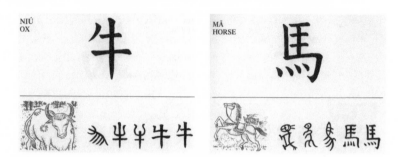

NIÚ
OX

MǍ
HORSE

Evolution from pictograph to ideographs to the modern
Chinese character for cow (left) and horse (right)

nese, the character for "cow" placed within the character for "strength" means "great strength." On the other hand, if you put the symbol for "horse" with the character for "strength," you do not get "great strength"—you get "brute strength," that is, just main strength and awkwardness. The horse has an even lower esteem in Japan, where the borrowed Chinese character for horse appears in the nouns for "fool" and "idiot."

In Japan, cows are the very symbol of things pastoral. In Japanese, the word for "pasture" has the Chinese cow character in it, as does the Japanese word for "bucolic" and also the word for poems of a pastoral or rural subject matter. In Western cultures as well, the cow is a rural icon. European and North American landscape paintings commonly use a cow, even if it is only a very distant cow, to symbolize the peaceful and restful quality of the scene. American paintings of wilderness, or unaltered nature, are characterized by an absence of cows, but depictions of the most rustic settlements or the camps of the earliest emigrants to the western United States have a cow painted into the landscape.

When Christianity arrived in Japan and there was a need to make up a word for a pastor, naturally they turned to the idea (and the ideograph) of cow and used the character for cow as part of a new made-up word for a Christian minister. In one of the most popular Japanese-language Bibles, published by the Japan Bible Society in 1963, in Psalm 23 the Lord is not a pastor—the word literally means "shepherd" and there is a Japanese word for sheep herder—he is a cattle herdsman, signified in a compound character that includes the symbolic cow. Certainly the beauty of the landscape in Psalm 23 is better imagined with a few serene cows by the still waters than with a bumptious gaggle of sheep chewing the green pastures down to the ground.

.   .   .

One indication of how much human beings have treasured cattle is the enormous variety of breeds. Like most domesticated animals, cows can be extremely variable in size, shape, and color, not only from generation to generation but within a single family. When exceptionally good animals appeared in a herd, they would be bred forward; the lesser animals would be eliminated. The outward and visible signs of essential excellence would be preserved—coat color, hornlessness, or a particular size and shape of horn. Cattle became identified with their valley or shire—red and white ones here, black and white there, and blond and gray and black. They became a source of local pride.

Even without including the hundreds of minor breeds, just sticking with the commercially viable modern animals, cattle are wondrously different from breed to breed. At a big state fair in the United States, visitors encounter a kaleidoscope of cattle: solid colors and blotched, some small, some huge, cows for milk, cows for meat. At a good fair there are more kinds of cows than of horses or pigs. About the only farm animal that can outnumber them in taxa is the chicken, and then only because of the amateur chicken enthusiasts who show their Bantams and

Cochins and so forth. On production farms, there are only a few kinds of chickens, and they are extremely boring in appearance and behavior. Cattle remain our most variable and specialized large domesticate exactly because they are our oldest and most valuable, tuned over the centuries to meet human needs in varied climates and landscapes.

And of all cattle at the fair, there are two breeds that never fail to attract favorable attention. On the milking side of the show barn, the delicate (for cows) Jerseys win the prize for beauty and amicability. Jerseys are a warm brown, darker on top, lighter underneath, exactly like other handsome quadrupeds—deer and elk. The most famous Jersey in the United States was Elsie the Borden Cow®, a trademark of a Jersey with a lei of daisies around her neck. She was such a celebrity that when shown at the New York World's Fair in 1939 and 1940, she outdrew every other exhibit. Borden's condensed milk, even before Elsie, was a staple food on the Western plains and simply called "canned cow."

In the beef barn, the Aberdeen-Angus, with their sleek black coats and their smooth hornless heads, are the very picture of a symmetrical package of meat. Exhibitors at fairs always put up a sign that directly, or in combination with the name of the ranch or dairy that is showing the animals, identifies the breed. For most visitors, the only one they already know is the Angus (in America) or the Aberdeen-Angus (to use its proper and original name) elsewhere. This is because the solid black, hornless silhouette of the Aberdeen-Angus steer appears on packages of meat in every major supermarket in the Western world and on the menus of countless thousands of steak-house restaurants. There are bigger beef animals with more bulgy muscles in their legs, but nothing beats the Aberdeen-Angus for elegance, as one might speak of an elegant solution to a mathematical problem. There is nothing superfluous about them. They are fine-boned but sturdy, fully muscled everywhere and still congruous. To use a parallel to what an English clergyman said of strawberries, comparing them to other berries, there is no doubt God could make a better beef animal, and there is no doubt that He has not.

The whole concept of breeds of cattle, as opposed to simply regional differences, is quite recent. Britons in the seventeenth and eighteenth centuries understood that some districts had unique animals. Durham in England had rather large red cattle; Hereford had red and white cattle; dark cattle came down from Scotland, some with horns, some without. But developing a systematic description of cattle happened only in the late eighteenth century, during the Age of the Enlightenment. The period may have been aptly named regarding mathematicians and philosophers and scientists, but for the average "enlightened" person, it really was the "Age of Naming Things and Putting Them in Encyclopedias." Not surprisingly, the end of the eighteenth century saw the first encyclopedic reference books for the various breeds of cattle. In an era of fierce nationalism, the authors wrote almost exclusively about the cattle of their own country.

The early English authors on domestic cattle, writing before there was such a science as paleontology and only the first stirrings of geology, always tried to take the history of cattle back to the remotest ori-

A typical Angus bull

gins, usually by mixing revealed truth with personal beliefs. Before Charles Darwin posited a gradual evolution and Gregor Mendel provided the first inklings of how variation might be transmitted to offspring, writers naturally innocent of the truth managed to settle the question in a sentence or two. The Reverend John Storer, an English breeder and judge of English Shorthorn cattle, had no difficulty naming the original locus of domestic cattle: "The native country of the ox, reckoning from the time of the flood, was the plains of Ararat, he was a domesticated animal when he issued from the ark." And then Storer added, with much metaphorical accuracy, that the ox "was found wherever the sons of Noah migrated, for he was necessary to the existence of man." It was also written that polled cattle like the hornless cows of Angus and Aberdeen existed from far beyond human memory. As a sort of evidence of that, a nineteenth-century Scottish farmer insisted that when the animals went on the ark two by two, there were four cows. One pair had horns; the other was polled.

Theological (or geological) assumptions aside, the biblical location is probably not far off the geographical mark. Once postdiluvian cattle realized they were well above tree line on Mount Ararat, a logical route downhill and downstream would bring them into the Tigris River valley and the rich land of the Kingdom of Babylon in Mesopotamia (modern Iraq). People who do not accept the biblical explanation also regard the Near East as the likely place of origin for the first domesticated cattle, as well as cultivated cereal crops, written languages, and urban centers. The fossil record and the archaeological record indicate that Mesopotamia was the epicenter of the agricultural revolution, or, as previously argued, the Cow Age.

But exactly where it all began is not as clear with cattle as it is with domesticated sheep and goats. The latter were each derived from just one species of wild animals with very limited geographic distribution. They were domesticated in an identifiable place and subsequently emigrated through most of the world. The aurochs, *Bos primigenius,* posited as the wild ancestor of the domestic cow, existed all the way from the Middle East and northern Africa to the island of Great Britain. Recent studies of the DNA of aurochs (from a few well-preserved bones) suggest considerable diversity in the extinct animals, and very likely there were several domestications from this variable ancestral stock. Put another way, it is entirely possible (although not necessary) to conjecture that in a time sequence that parallels that of biblical chronology, the isolated British cattle were a separate breed from the pair(s) on the ark.

William Youatt, the early-nineteenth-century author of the eponymous reference to the British breeds, *Youatt on Cattle,* remarks on the cow's earliest, antediluvian history by writing that "Jubal . . . who was probably born in the lifetime of Adam, was the father of such as have cattle." Youatt (or his printer) erred: It was Jabal, Jubal's half brother, who "was the father of such as dwell in tents and of such as have cattle." Brother Jubal "was the father of all such as handle the harp." They were both great-great-great-great-grandsons of Adam, which would make Adam rather elderly at their birth. And he was; Adam lived, according to the Book of Genesis, to be 1,060 years of age, making him, as the expression goes, "older than Methuselah" (who died at 969).

To regard the relationship of Jabal, or Noah, and cattle as speculative is not to make fun of the eighteenth-century authors. Humans try to come to inquiries with an open mind, but they cannot bring a truly blank slate. Authors in the post-Darwin era also tended to read their personal prejudices into the putative history of British cattle, and the Aberdeen-Angus cow in particular. One theory of the origin of the small horn-less black cows came from Richard Lydekker, who wrote *The Ox and Its Kindred* in the first decade of the twentieth century, a time when concern about human racial characteristics abounded. Then was the flooding tide of the eugenics (good genes) movement, which advocated everything from sterilizing the incompetents and malcontents to severe restrictions on immigration of even slightly brown people into Anglo-Saxon or Nordic, that is, white, societies.

Referring to all the highland cattle, horned and polled, Lydekker asserted that they were

> *introduced into the British Isles by the small dark Iberic race, now mainly to be found in the western parts of our Isles, in Wales, Scotland and Ireland [he means the human beings, not the cows] and [the small dark Iberic race is] still to be recognized elsewhere in our population by the small dark folk. [Their] . . . cattle were small and dark, with small horns, and were the only domestic breed in this country, so far as I know, throughout the whole of the Bronze and Iron Ages, and during the time when Britain formed a part of the Roman Empire.*

The counterpart to these undersized, overpigmented im-ported cattle would be, to Lydekker, the so-called White Park cattle of Britain, presumed to be preserved in estate hunting parks, and representing the authentic British ox. "These small cattle . . . ," Lydekker added, meaning the Celtic imports, "still live as the small dark breeds of Scotland, Ireland, and Wales. This breed contrasts in every particular with the large white cattle" found in the hunting parks of British nobles. In one re-spect, Lydekker was correct, even if accidentally and condescendingly so. The Celts (for whom he uses the term Iberic), as noted, were always cattlemen.

In case a reader is personally acquainted with the current breed of Scottish Highland cattle and wonders why Lydekker calls them a "dark breed," it is because they were basically black on brown. The paler modern animals often encountered on farms and in public parks were selected and preserved from the utterly diminished population found in the nineteenth-century highlands. Once again, as with all domesticated animals, there is genetic variation in color that can be exploited by the breeder.

Another possible source for most of the "small dark breeds" could be the various Scandinavian settlers. Herds of ancient, and for the most part no longer present, breeds with a common characteristic were once found along the coasts of Scotland, Wales, and Ireland and the northern isles, and as far as Iceland—that is, everywhere the Danes had invaded and founded settlements. These cows were frequently polled. Not all were hornless, but there was a tendency toward polledness. Where they were closely herded and bred to purpose—as in Galloway, Angus, and Aberdeen—the native animals were almost entirely polled. This connection between the Baltic raiders and hornless cattle makes sense. In spite of the helmets worn by fans of the Minnesota Vikings football team with those great sweeping Wagner-inspired horns, it seems more reasonable to think that the Danemen and Norsemen preferred bringing hornless cows on their voyages. Imagine one of their cramped long, lean ships heaving on the North Sea waves, and imagine a horned bull lashing about himself. No, the Norsemen were very bold and very brave, but they were not stupid. The great Viking drinking horns were like the German ones described by Julius Caesar, and were also aurochs horns. Examples of silver-decorated aurochs horns associated

Scottish Highland cow

with Danish artifacts have been unearthed in British bogs, nicely preserved. If the Baltic invaders introduced their polled cattle, they were not the first. Archaeological remnants of the eighth-century Danish settlements discovered in modern Dublin, Ireland—the very oldest Danish settlement in Ireland—lie above skulls of polled cattle that had been deposited long before the Vikings arrived. This suggests a preference for hornless cattle well back into the era when the Gauls (Celts) dominated all of western Europe.

The English Park cattle that Lydekker presents as the aesthetic opposite and superior to the dark runts brought by the Celts got their name because they were kept in baronial hunting parks along with red and roe deer and other objects of the chase. The Park herds represented to their admirers the virtues of the Anglo-Saxon race. Park animals existed from time immemorial, had sprung from Albion's land, and were white.

In his remarkable treatise *The Wild White Cattle of Great Britain*, Reverend Storer argued that the Park cattle still surviving in a few baronial hunting preserves were wild in behavior and descended directly from some indigenous British beast. Park cattle were regarded as, literally, fair game. They were hunted like the stag, although rather rudely in times past. The annual hunt at Chartley Castle's hunting park was a sort of free-for-all, with all the neighbors invited in to surround and shoot at a designated bull with whatever kind of firearm they owned. Since most countrymen owned shotguns, and most had perched themselves in trees, just in case, the amount of lead scattered around Chartley was considerable. Festive as it was, it was too dangerous and perhaps too riotous and democratic for the Victorian era, and the hunt changed. Bulls were killed at Chartley in the late nineteenth century by a single marksman. The last of all was shot by the Prince of Wales, and it was duly noted in the press that his highness's marksmanship was exemplary.

Deeming the Park cattle wild (and one version survives as a very minor boutique animal, not surprisingly called White Park cattle, noted for their good behavior and good foraging habits) was a fantasy from the beginning. Storer himself provides enough information on the animals and their care by the parks' herdsmen to undermine his own thesis. Storer, who clearly understood the role of selective breeding in cattle, reported that they were highly variable within a Park and between Parks. There were white cattle with red inside their ears that had the occasional calf with black ears instead, and a Park herd with white cows and black ears would occasionally drop a red-eared calf. In both cases, the atypical animals were destroyed. Solid black calves were not uncommon, and they were consigned to the shambles on the spot. There were also several herds of polled Park cattle, which Darwin noted in his *Variation of Plants and Animals under Domestication*. All these myriad variations are only typical of domesticated ungulates and unheard of in wild ones. The uniform configuration of each Park cattle depended entirely on the efforts of the herdsmen to weed out the nonconformers. Everyone realized what was going on, but, on the other hand, it wouldn't be sporting to shoot one if the landlord and his peers didn't declare it wild.

The Reverend Storer was a Shorthorn man, like all good Englishmen of his time, and a judge of beef cattle in English fairs. While he promoted the notion that the Park cattle were somehow aboriginal, he did acknowledge their well-developed qualities. He sounds almost exactly like a Scot praising Aberdeen-Angus: "Straightness of the back and belly-lines was a most strongly marked feature of these cattle; depth also of the body, and shortness of leg." He remarked, somewhat wistfully, that the "wild" cattle of Chartley had exactly the characteristics so desired, and so far unattained, by the Shorthorn breeders. That they came by these qualities from the rigorous selections made by the herdsmen is not something he could admit. Here indeed was strong evidence of variation by selection of desired traits, but the Reverend Storer, with his faith in the story of Noah's ark, was eager to believe in the authenticity of the Park cattle, even though he was well aware that, like the Shorthorns, they were a breed developed by herdsmen. Another clue, amounting to proof, that the White Park cattle had been selected by the herdsmen was still hidden from everyone's sight. The great cave paintings of Europe awaited discovery, with their naturalistic representations of the ancestral aurochs as a very dark (as well as humongous) beast.

·   ·   ·

Darwin himself was hardly sure of what was actually happening to cause these variations within and among the British herds of "wild" and domesticated cattle. That is not a criticism, just a statement reflecting the world's understanding of heredity throughout the nineteenth century. A close reading of volume I of his work on domesticated animals—*The Variation of Animals and Plants under Domestication*—suggests that he thought environmental influences produced changes that, when selected for by the breeder, became heritable. That is, the environment did not always just act upon random variation, rewarding the good and punishing the bad: It could *cause* the variation. He described several instances, including British hunting dogs, where when the animals were taken to a tropical climate they immediately produced offspring entirely dissimilar from the parent stock, a result he attributed to the local heat and humidity. This would make him a Lamarckian. This is the theory of Jean-Baptiste Pierre Antoine de Monet, a/k/a the Chevalier de Lamarck. The French philosopher/naturalist, who died in 1829, two years before Darwin's momentous voyage on HMS *Beagle,* posited that specialization was caused directly, if incrementally, by the environment and behavior of the animal. To Lamarck, the elongated neck of the giraffe was *not* caused by minute and random variations being favored over the eons by a critical Mother Nature: The variations were caused by the parent animals stretching to the utmost, generation after generation, and altering by such exercise the nature of their sperm and ova and embryos. Lamarckian biology survived in the work of the greatly misguided twentieth-century Soviet geneticist T. D. Lysenko, long after the rest of the world (and most of the rest of the Soviet Union) had adopted a modern view of variation and heredity. Lysenko didn't have to defend his atavistic biology, for he had an enthusiastic supporter, Joseph Stalin.

Given the mystery of how heredity actually operated, Darwin could only speculate that "some peculiarities, such as being hornless, &c., have appeared suddenly, owing to what we may call in our ignorance *spontaneous variation.*" (Italics added.) It is that phrase that evolutionary biologists emphasize while placing Darwin in the pantheon of science. He certainly deserves to be there for making an irrefutable case for variation over time and for selection of traits by a punishing nature or a nurturing agriculturalist. He put it quite differently when he spoke of polledness and other mysterious changes than when he wrote of the influence of the environment (including the state of domestication) on the variation of animals. Only the inexplicable and peculiar variations that *cannot* be attributed to environmental conditions or the life history of the organism Darwin called "spontaneous variation." It was just one of many causes of change for Darwin, not the unitary theory of today's genetic explanation of evolution. His spontaneity is what we would call a mutation and, in particular, a mutation in either a spermatazoa or the nucleus of an ova—a germ-line mutation. In a letter to the authors of the first history of Aberdeen-Angus cattle (Mssrs. James Macdonald and James Sinclair), he remarked that while "no one can give an explanation—although no doubt there must be a cause—of the loss of horns. . . . It is, I think, possible that the loss of horns has occurred often since cattle were domesticated, though I can call to mind only a case in Paraguay, about a century ago." When he writes of "cause" the reader gets the full effect of biological speculation in an era of very primitive understanding of inherited traits and, in particular, of the effect of dominant and recessive genes. Darwin's remarks that "no doubt there must be a cause" implies that there must be a larger plan, a purpose.

Darwin lived a country gentlemen's life and was particularly fond of breeding pigeons and acquiring rare strains of the bird. He was entirely cognizant of all of England's domesticated animals and of their variations. What would have struck him particularly about polled cattle is that such a fundamental structural change was not accompanied by any negative consequences. After all, a hornless cow is as much a freak of nature as a horse without a tail.

As a farmer (with hired help to do the heavy lifting) and a correspondent with hundreds of individual agriculturalists and associations, Darwin was familiar with spontaneous curiosities being born or grown from seed, usually with fatal consequences. It was known among people who raised goats, for one example, that when a horned breed suddenly produced a hornless billy or nanny, the animal would almost certainly be functionally sterile.

Today we understand what went wrong with these polled goats because we know about genes, the one great gap in Darwin's understanding of the world. Polled goats, and their associated problems, are classic examples of multiple consequences of a change in an animal's genes. A single genetic mutation may have two effects or more, and often quite negative consequences for the organism. Having multiple outcomes from a single mutation is called pleiotropy, and it is a very good thing that polledness in cattle is *not* pleiotropic. For example, when horned breeds of goats get a mutation for hornlessness, it is recessive (that is, it must be on the chromosomes contributed by both male and female parents to do its work). That's the good news, but when the females get the polledness gene from both parents, and thus have a copy on both chromosomes, they are not only hornless but also develop as intersexes. Boy goats with two copies of the polledness gene are often sterile. This pleiotropy keeps polledness (in the usually horned breeds of goats) from being terribly widespread, as the intersexed females do not reproduce, and many of the polled males cannot breed and thus continue supplying the population with a copy of the recessive gene for polledness. The pleiotropic gene does survive, however, exactly because it is recessive, just as some inevitably fatal human genetic diseases, Huntington's chorea, for one, survive because there is a reservoir of single-copy humans.

The spontaneous loss of horns in cattle has occurred often since domestication. The absence of horns in cattle, we now understand, is caused by a variation that creates a dominant gene, actually a pair of dominant genes that live next door to each other on the cow's chromosome number one. In a few years, somebody will have sequenced the pair and described exactly how it happens that the gene for horns turns into the gene for no horns. It could be that a piece of DNA flips end for end, drops an element, or just mixes up its sequence of amino acids. From its relative frequency and power, it will turn out to be a simple and elegant alteration in the natural genetic state of cattle. By some good luck, polledness in cattle has no effect on any other aspect of the animal's growth or conformation or behavior. Polled Herefords, a twentieth-century innovation developed by breeders, are otherwise identical to their more widespread horned cousins and just as valuable, if not at a premium, because they are less likely to make trouble.

Although hornlessness is entirely accidental, until the herdsman starts to select for it, it would appear to prescientific farmers as a kind of blessing. It would be the stockman's equivalent of a "blessed event." The polled trait was seized upon from Egypt to Ireland. The Scots, from the earliest times, have valued their black polled cattle. More than one author has noted their affection for their hornless "coos," a statement that sounds odd when the long-term point of raising Aberdeen-Angus is to eat them. But cattlemen worship at a different altar from dog owners and cat cohabitors.

· · ·

One sign of the kind of affection bestowed on their black hornless cows is the nicknames Scots gave them long before there was a city of Aberdeen or a county of Angus, centuries before Scots decided to call their special animal the Aberdeen-Angus. Nicknames are always a sign of congenial familiarity. Persons from anywhere in Scotland would simply call the ancestors of today's Aberdeen-Angus "black sleekies." That distinguished them from the equally black and polled but curly-coated Galloways. Besides black sleekie, there are two provincial sobriquets.

In southeastern Scotland, in the former shire of Angus, they are called dodded cattle or, more often and more intimately, doddies. The addition of the ending, from "dodded" to "doddie" and "sleek" to "sleekie," connotes the same familiarity as when used on human beings, when grandmother becomes grammy; momma, mammie; or when the Scotch-Irish of the Appalachian Mountains turn papa into pappy. (In Mark Twain's *Huckleberry Finn,* Huck calls his despicable father just plain "Pap.") Other breeds might have descriptive nicknames, as Herefords are "white faces," but they are never called "herfies." The affectionate sobriquet was unique to the hornless black animals.

The source of the word "dodded" is utterly mysterious. It may be applied throughout both England and Scotland as an adjective for a tree that has been deliberately pruned back to a stump, that is, a pollarded tree intended to grow many small straight branches that burst skyward from the trunk. Most commonly pollarded were the beech trees (everything from tool handles to bentwood furniture could be made from the numerous branches) and willows, whose many small branches were used in basket weaving. As the dodded tree is regularly trimmed back for new branches, it develops a large lump at the point of the pollard. With horns in a state of utter deficiency, the most noticeable thing about an Angus head is the often-prominent rounded boss on the very top of their heads between the ears. Another generic name for that bump is simply poll. Since the poll is so obvious, not obscured by horns, it became a verbal adjective, "polled," shorthand for hornlessness.

The Angus district use of this word for polled cattle seems to be unique to that area of Scotland. "Dodded" also has a local meaning up in Aberdeen, not for the Aberdeen district's cows but for another peculiarly Scotch object. The top of a whisky still's chimney is a large rounded dome that caps the chimney. It somewhat resembles a squished version of the "onion" domes of Byzantium, most familiarly seen upon the Kremlin in Moscow. Along the banks of the Spey in Aberdeen, this chimney cover is called a kiln-dodd.

In the northeastern counties and districts, Banff and Aberdeen, Forfar and Moray, and up into the Hebrides, the locals called the polled black cattle hum(i)lits and hummels, almost always familiarized as hummlies, following the same course as Angus's "dodded" to "doddie." Many have assumed that this was dialectical for "humble." Dr. Johnson, in his dictionary, was the first perpetrator of that explanation: "Of their black cattle," he wrote, "some are without horns, called by the Scots *humble cows*, as we call a bee an *humble* bee that wants a sting."

Humble bee is British for what most Americans call bumblebee, the large black and yellow bees that do sort of bumble around from flower to flower, bending the stalks of their commonest nectar plants, the various clovers. They have an extremely functional sting and quantities of venom. Dr. Johnson apparently did not spend much time walking barefoot through the fields. Bumblebees are just slower to anger than the average hymenopteran. The other kind of "humble" bee in English (but *not* Scottish) dialect is the drone of the honeybee tribe, and drones do lack a stinger.

In Scotland, there is never written or spoken any suggestion that polled Aberdeens or Angus were somehow humbler simply because they were sweeter-tempered and horn-free. If the source is the Latin-based English word "humble," there is no ancient record of its ever having been spelled that way. Dr. Johnson attributed the mispronounced (in his opinion) "hummel/ hummilit/hummlie" to local, ignorant, dialectical speech.

To the author's knowledge, the fourteenth-century archives of Elgin, a city midway between Aberdeen and Inverness, contain the first written reference to polled Scotch cattle. The record describes: "Ane cow of four zeiris [years] auld with her calf black hummilit was sauld in Quhytwer [Whiterow] for sevin marks." Fans of Robbie Burns will note that the dialectical orthography of "one" and "old" and "sold" from ancient Elgin is identical to his. A sixteenth-century legal document from Aberdeen records in Latin the transfer of a piece of land and *"unum bovem nigrum hommyll appretiatum ad quadragintas solidos et octo denarios monete Scotie."* That is, "one black hommyll cow valued at forty shillings, eight pence, in Scotch money." The only word not translated into carefully spelled Latin is hommyll. For consistent Latinization, if the term "hommyll" was understood as a dialectical form of "humble," the lawyer would surely have written the similar, but really quite different, Latin for "humble," *humilus,* not "hommyll." Note that shillings and pence became *solidos* and *denarios,* so eager was the lawyer to make every possible word a Latin one.

The likeliest source of the Scottish dialectical "hummlie" is not Latin but that other great mother of English as we know it, any and all of the old North German languages. It could have arrived with the Angles and the Saxons in the sixth century A.D., or it could have been transported a few centuries later by the Norsemen, whose language derives from Old Saxon. It may have arrived on a Danish longship with the animal of the same name. The Old Low German "hummel" or "hommel" meant a hornless beast of any normally horned or antlered species, and it would be used in combination: a "hommelroh," for example, would be a male roe deer that lacked horns. It survives in Bavarian dialect to this day as a combinative/adjective—"humlet," almost identical to that earliest reference from Elgin, "hummilit." Some of Dr. Johnson's reference still graces the entry in the latest edition of the *Oxford English Dictionary,* but the part about humble bees has been expunged. On the other hand, Dr. Johnson might have been accidentally correct about the etymological (and entomological) relationship between hummlie and humble bees, if wrong about the humbleness. In southern Germany and Austria, variants of hommel/hummel are used to name the big solitary bees known to Johnson as humblebees and to Americans as bumblebees.

And the old German/Saxon/Norse word also survives in England in that most archaic of specialized jargons, terms of venery—the argot of the hunt, most of which is made up of French terms introduced by William the Conqueror, who also introduced Norman laws regarding who owned the wild animals. (Basically, the king owned the animals and would parcel out ownership to loyal barons, earls, sheriffs, and other sycophants. This is the basis for all the goings-on about deer hunting in *Robin Hood.*) A naturally polled stag, one that should have but will never have antlers, is a hommel stag, a quarry unpopular with huntsmen and landlords alike because the trophy is absent.

But to the Scot, his polled black beasts were so very far from humble that they constituted a kind of animal royalty. One of the great matriarchs of the Aberdeen-Angus breed is Pride of Aberdeen, and there are any number of bull and cow families with Prince, Queen, Judge, and similar honorifics. Indeed, one of the intangibles of judging Aberdeen-Angus cattle in the showring is their aristocratic bearing: Bulls should be lordly;

cows and heifers feminine; but to be placed high, they must all have an aura, something bordering on a personality trait like self-confidence or self-assurance, of being to the manner born.

Bill Sklater, an Orkney man, or, as they say there, an Orcadian, was the author's guide on a tour of Aberdeen Angus herds in northeast Scotland. He has been a judge of the breed since the 1950s, and today he helps farmers pick out breeding stock for their farms. In short, he is an expert; one might call him a cow consultant. As we traveled from farm to farm, he remarked on more than one occasion that an Aberdeen-Angus cow or bull in a pasture "had style." It was a high compliment. What he meant, it became clear after the fourth or fifth stroll through a paddock, was that the animal that caught his eye stood as if it *knew* that it was indeed a stylish beast. It was not just a question of size or conformation in the eye of the beholder; it was a question of justifiable self-esteem on the part of the animal. Put another way, the animal was charming, attractive, as close to having a personality as a cow can get. The affectionate nicknames for the Aberdeen-Angus become more understandable when a visitor sees them in small groups on their native soil. "I often find," Sklater said, speaking of this stylishness, "that the first impression, the first sight you have of the animal in the showring, is the best judgment. I have second-guessed myself many times, and thought afterward that the first impression was the correct one."

That certain ineffable aura that a prize animal has is not easily described. William M'Combie (1805–1880), the first great international showman of the breed, said much the same as Sklater about an animal's appearance: "A perfect breeding or feeding animal should have a fine expression of countenance— I could point it out, but it is difficult to describe upon paper. It should be mild, serene, and expressive." A prominent American breeder, James B. Lingle, manager of the Wye Plantation herd in Maryland, when writing about the foundations of his herd, described his ideal cows as large, rangy, and coarse, with big heads and wide jaws. Wye cattle were always bred for size with little attention to elegance. Although he preferred big ungainly females, even Lingle could be charmed by a feminine cow: He described the only averaged-sized dam in the Wye Plantation's foundation herd this way: "She appeared a bit delicate-headed, but her eye was good, she had a trim neck, and she was indeed a fair sort of lady cow." That is not a redundancy; "lady" is a complimentary adjective, as in ladylike.

The standard of the Aberdeen-Angus breed, the instructions to the judge, and the goals set for the breeder always include a category that moves from the physical traits on to something like personality. The Canadian standard is typical: The general appearance of a bull should be "Alert, well-bred, masculine." Females should be "Gay, well-bred, and feminine." Note that the adjective "well-bred" is not about genetics or parentage; it is an old-fashioned and now politically incorrect description of a person's good manners and polite behavior, an adjective for self-confidence without braggadocio, courtesy without humility. The breeding standard has excruciatingly detailed instructions for judging the body of the beast, but just those three significant words about the aura.

It is true that the mien of most cattle, most of the time, is one of drool, stool, and apathy. But a haltered cow led into a show-ring, coming into the glare of the lights and under the stares of strangers, can be very alert and quite gay. It may be difficult for someone who has never been to a Scottish cattle show to imagine a black cow singing Maria's song from *West Side Story,* but indeed, standing in the showring, her hair brushed and her hooves shined, she does feel pretty, oh, so pretty.

→ 3 ←

A Unique British Cattle Culture Develops and Successfully Emigrates to America

AS the Scottish names for their favorite cattle are ancient and well established by the time of written records, so too is another element in the history of the Aberdeen-Angus cow. The Celtic Britons—that is, the Scots and Welsh, and to a lesser degree the Irish—developed a style of raising their many kinds of cattle that took advantage of the realities of their landscape. The Celts occupied the difficult landscape of Britain, the narrow valleys and high country of Wales in the west and Scotland in the north. Their adaptation to this rugged land created a pattern that is still in use, not in Scotland today but in the North American West.

In short, they reared cattle by feeding at a home base through the inclement winter, taking them to summer pastures on unimproved rangeland or mountain slopes in summer, and selling lean cattle to farmers with richer pastures. Before the nineteenth century, cattle from Scotland, Wales, and Ireland went south and east to England, where they were "finished" for slaughter. When emigrant Scots and Irish came to America, they carried this pattern with them. It survives to-day in the American West, except that railroads have replaced the cattle drives and feedlots the rich pastureland of England. The system flourished in Britain until a revolution in transportation and horticulture in the early nineteenth century led to the current British practice of fattening on farms and shipping finished live animals to market. Those changes, occurring first in agriculture and then in transportation, allowed the development of Scotland's purebred cattle.

From at least the era of the Roman occupation, Celtic Britons (referred to earlier as Lydekker's "small dark folk") were raising cattle and driving them down from the hill shires of Scotland and Wales as lean cattle and selling them to be fattened on the lowland pastures of England proper—then to be driven on to markets in the great cities. Black polled cattle from Scotland, the ones that would be regularized as Aberdeen-Angus and Galloways in the nineteenth century, were driven with other lesser breeds across the borderlands and down to the richer, milder, southern pastures. In one historian's shorthand, it was "Celtic breeders and Saxon feeders." The same pattern prevailed in Ireland, with Irish cattle finding their markets after being shipped across the narrows of the Irish Sea from northeastern Erin to ports on the west of Scotland, thence to be driven to market. Historic cattle trails, many of them fenced with stone for leagues, flowed south, joining like small streams into a great river of cattle destined, once fattened, for London and the other bustling ports and cathedral cities of England.

This trade in lean cattle had become so standard a practice by 1500 A.D. that there were professional drovers who carried whips (the original "bull whips" but with a longer handle and a shorter cord than the ones used by villains in American movies) and who owned herding dogs (of no particular breed, unlike the shepherds who had already created their various regional collie dogs). It was a trade as recognized as blacksmithing or weaving or carpentering. (Given the Scottish tradition of clan names as surnames, "Driver" is not a terribly common last name in the English-speaking world compared to Carpenter, Weaver, or Smith.) As England's empire grew and the mercantile cities and the first stirrings of industrialization began in The Midlands, the stream of cattle coming south became a torrent.

Centuries of droving left their mark on Scotland, but it is increasingly difficult to find the traces today. The drove roads were fenced with stone, and the rut created by thousands of cattle over hundreds of years enhanced—by reduction—the height of the stone walls. Where the drove trails lay on practical routes from one village to another, they were replaced by modern roads, from the first graveled cart roads on to the modern asphalt highways. The drove roads over mountain barriers, from glen to glen, may still be there, but walking about is not much encouraged by the great landowners of Scotland.

A drove road

Driving on the A9, the north-south express highway from Perth, Scotland, to Carlisle, England, a motorist turning east toward Falkirk (via Muckhart) on a good country highway, the A823, will see across the glen and well above the wee burn a wonderfully preserved (and private) section of the drove road that ran from Falkirk west to Blackford, where it joined the main flow south to Carlisle. This section of ancient road is conveniently close to one of Scotland's major tourist destinations—the Gleneagles Hotel and golf courses. The steepness of the glen's sides has two advantages. For the tourist, the drove road is easily visible a short distance away across the narrow glen, where it runs at about the same elevation as the A823. For the drovers, the steep pitch of the hillside meant that they had only to fence the downhill, north, side of the trail; the glen's own precipitous slope made a natural wall on the opposite side of the drove road.

Carlisle, a cathedral town in extreme northern England, became the major port of entry for cattle. Carlisle has always been a place where Scots descended on the English, sometimes with cows and sometimes with claymores. The cathedral there resembles a fort, built much more stoutly than its Gothic counterparts in southern England, complete with arrow-firing slits at the ground-floor level instead of windows. Carlisle sits just south of Hadrian's Wall, built anew to keep Scots in their place in the third century A.D. For most of its length, the wall follows along the northern edge of steep natural escarpments, but at Carlisle the ground softens down to a rolling-hilled plain, a natural passageway for all the Scotch cattle headed south to The Midlands. By the end of the seventeenth century, beasts by the tens of thousands annually crossed from Scotland into England at Carlisle. And cattle, archaeological investigations suggest, had moved south through Carlisle even before Hadrian's Wall went up.

Lean cattle, in the general sense of the phrase, are what Westerners raise and then sell to feed yards, where they are fattened (excessively, some think) before slaughter. The animals driven south from Scotland were surely leaner than lean. No great care had been taken in breeding them, and no particular effort had been made to provide them with a richer diet than wild grasses and forbs and the straw left from threshed grains. What makes this Scot business particularly relevant to cattle ranching in Montana or anywhere in North America is that the cows were not left to fend for themselves through the entire year. Their winter diet was scanty, but it was stored for their benefit and provided on a daily basis. They were cared for like part of the family by their owners, a practice that gives us that otherwise mysterious phrase "animal husbandry." Dairymen throughout Europe fed their milk cows through the winter (and took them to mountain pastures in the summer). Continental farmers each fattened their own beef for the local market. The uniqueness of the Highland system was the division of the beef industry into two parts, two stages of ownership: Cattle would be bred and ranged in the inhospitable high country and then readied for market by specialists who would fatten and condition them in the richer, more temperate farmlands of England. For centuries, in the cold damp air of Scotland, farmers could grow only enough fodder to keep the breeding animals through the winter; the surplus cattle had to be fed elsewhere.

Scottish beef cattle were typically driven into the hills in spring, up to an area of moderate vegetation between the lowland farms and the barren and not very nutritious heath-covered moors on the high ground above the tree line (which was the grass, herb, and shrub line as well). Herded down in the fall, some would be culled for market, the rest kept through the long dark winter. This is exactly how most beef cattle are raised in the North American mountainous states. They are taken to unimproved summer pastures above or away from the irrigated fields, and then fed through the winter on a boring but nutritious diet of hay and water and given some mineral supplements, which are not much more than scientific salt licks. The practice was so common in Celtic Britain that there's even a Gaelic noun specific to a summer cow camp in the hills; it is a *shieling*.

In those days the distinction between a milk cow and a designated beef-style calf-raising cow was not a question of breed; it was merely a matter of individual differences within the herd. What the Highland Celts were selling to market before the agricultural and transport revolution of the nineteenth century were not in any modern sense beef cattle. Even when the first registrations of Aberdeen-Angus were made in the 1820s, many of the purebred cows were being milked for home consumption. It may seem impossible for a cow to suckle a calf and be milked at the same time, but a good Aberdeen-Angus mother has the capacity to raise twins. If she has but one calf (which is by far the most common situation), the surplus can go into the farmer's bucket. One of the founding mothers of the breed was bought specifically as a milk cow; her beef potential—when her first calf grew into an impressive meaty animal—was a pleasant surprise. The British Shorthorns (the first cattle brought to the Americas to improve herds) were also a "dual-purpose" breed until well into the nineteenth century, when they were first selectively bred as either "Milking Shorthorns" or just plain beef Shorthorns. The milking side of that family is still raised on small dairies in New England, but most often as a something-to-talk-about hobby herd on a dairy farm where Holstein-Friesians pay the bills.

While the black polled cattle of Scotland had always brought premium prices when sold in England for fattening, the serious work of breeding them specifically as meat animals and, more important, selling them as fat cattle and getting the premium price into the farmer's purse had to wait until the Scots, in their short cold summers, could raise enough food to not only keep animals alive through the winter but to improve them, to fatten them up to market quality. The crop that made winter feeding possible was the turnip. Not the delicate purple-top white turnip of the supermarket, but a sort of giant rutabaga, that strong-flavored huge—if allowed to grow all summer—yellow-fleshed root with a, well, memorable flavor. Using turnips and other cruciferous roots for cattle food was a continental invention. The Germans had their sugar-beet-like mangel-wurzels (and still do), while the Scots brought in rutabagas. They were (or it was thought they were) imported from Sweden to Scotland. When tourists look at the vegetable side-dishes on a Scottish menu, they will find something with the otherwise wholly inexplicable name Mashed Swedes. This is the culinary rutabaga, boiled to a fare-thee-well, pulverized, buttered, and piled on the dinner plate. Samuel Johnson is famously remembered for defining oats with a verbal sneer in his dictionary as a grain that is fed to horses in England, whereas in Scotland it sustains the people. The rutabaga, prized by cows and people alike, lends itself to no such English (Sassenach) effrontery.

The huge yellow turnip and the modern Aberdeen-Angus breed are historically inseparable. Not only did the introduction of this hardy and nutritious vegetable revolutionize cattle raising in the damp north, it had a very direct effect on the pace of exporting Highland patterns of animal husbandry to the American colonies.

The growth of capital-intensive farming by the major landowners required new farm machinery, draft animals, and seed for the new forage crops—turnips and alfalfa. All of this was beyond the reach of the tenant farmers, the crofters, already living on the edge of starvation. Farms increased in size and the tenants became employees, hired hands, and learned the intricacies of breeding cattle and raising the crops to feed them. This semi-industrialized cattle raising encouraged the last great migration of Scots to North America in the nineteenth century. Where the revolutions in Scotland against English rule in the eighteenth century and the enclosure of land by the victorious English had famously sent hundreds and thousands across the sea, the agricultural revolution that peaked at the beginning of the nineteenth century meant more enclosures for cattle pasture and turnip fields, thus causing another exodus. In addition to the changing use of the land (more fields meant fewer tenant farmers on an estate), there was a long period of depressed economy following the end of the Napoleonic Wars and the auxiliary War of 1812 in the United States. As with most depressions, it weeded out the small and the economically weak. The largest number of applicants for charitable funds to buy passage to America were farmers, followed by shopkeepers. Given the distribution of land in Scotland into vast estates, men calling themselves farmers were tenants not freeholders; many of those who emigrated in both centuries were cow-men (an older word than "cowboy").

When the nineteenth-century Scots arrived in the new United States with their informal education in the new agriculture, they found, for the most part, a familiar method of stock management. In the eighteenth century British emigrants had brought the Highland system to the American colonies. In particular, the arrival of the Scotch-Irish in the southern colonies that were expanding westward into and over the Ozark and Appalachian Mountains married two essentials: a lowland-highland geography and new settlers accustomed to exploiting that kind of landscape. The idea of turning the animals into the hills in clement season and then gathering them up and bringing them in to be fed through the winter is written on the map of the montane American South—more than a half-dozen places and streams named "Cowpen(s)."

.   .   .

The methods of the Scots (and Welsh and Irish) did not triumph everywhere in the Americas. British immigrants who moved into the Southwest in the early nineteenth century, to western Louisiana and eastern Texas, adopted the indigenous Hispanic style of cattle raising. It was a system without agriculture—no irrigation, no haying, no growing of feed. The animals were left on their own. This landscape had swamps, thickets, and grasslands, utterly without mountains or hills. The up-down, summer-winter, pattern of husbandry was both impossible and unnecessary. About the only change the Anglos made to what they found was to bring along the British style of brands—block letters and numerals and a few simple symbols. These quickly replaced the ornate Moorish patterns of intricate swirls so typical of Spanish and, by export, Mexican brands.

Hispanic *vaqueros* were essentially cow capturers, not animal husbanders. This style of ranching required short bursts of activity and offered long periods of near idleness. In the semitropical coastal plains between the Mississippi and the Rio Grande, this system worked well enough and the longhorns multiplied without man's assistance. The same system operated in Spanish California, with vast herds and minimal management, and in both places the major commercial trade was not in meat but in hides and tallow. This ranching style developed in the very temperate Spanish province of Andalusia, where wild and untillable marshes provided year-round forage for semidomesticated cattle. It is not theoretical that this southwestern Spanish province was the model for the southwestern United States. Andalusia was the great source of Spanish emigrants to the Americas, a fact easily discerned in the sound of the Spanish language from Mexico to Tierra del Fuego. In Andalusia, *S*s and soft *C*s are pronounced much as in English. In central Spain, around the old kingdom of Castile (and in Spanish schoolrooms and on national television), the letters *s* and *c* are lisped, aspirated. Castile comes out Cathtile and Andalusia becomes Andaluthia, etc. Americans who learned Spanish in school are usually taught the Castilian accent. It will make them no friends in Andalusia, where some small shopkeepers have signs saying the proprietor does not speak and does not understand Castilian Spanish. As most of the Spanish colonies had even more extensive wildlands and even more temperate climates, they found no more reason to change their Andalusian husbandry than to change their accent in the New World.

The Anglo cowboys of the Southwest learned cow catching from their Hispanic neighbors and mastered the skills necessary for that art, roping and riding. The archetypal cowboy in the movies (*Red River*) and television specials (*Lonesome Dove*) is always a Texan. This is accurate enough. Texans made the long drives; Texans first, and briefly, worked the big open-range ranches on the High Plains of Montana and Wyoming. Unfortunately for their investors, what the Texans brought north besides long-horned animals and lariats was the fairly shiftless ranching program they had learned in southern Texas.

The Spanish system's defining characteristic—leaving the cattle to range year-round—might best be called benign neglect. As it was applied in Hispanic-American Texas, the longhorns were half-wild, and they were expected to get through life with minimum human contact. The little human interactions they had were basic and unpleasant: castrating, branding, slaughtering. Herds mingled on the grazing grounds and were sorted out by cooperating stockmen. The calves were branded to match their mothers, and the steers and aging animals were culled for market. The word "roundup" wasn't commonly used in Texas; Anglos called it cow hunting. A group gathered to hunt cows was called a posse, from the Spanish phrase *en pos de,* meaning "going after." The New World Spanish for this business of hunting cows is *rodeo,* which moved into English as the word for an exhibition of cowboy skills required to capture free-running cattle. Its original meaning is another indication of how little in the way of physical structures were involved in the Texican cattle business. It comes from a verb, *rodear,* which means to surround something, as opposed to putting something into an established enclosure like a corral (another Spanish contribution to the language).

Lacking anything in the way of corrals and fences to separate cows from their calves when weaning time came, Texans weaned the calves by slitting their tongues. Another example of how little capital needed to be invested in Spanish or Texican ranching was the absolute absence of barns. When modern Spanish speakers have to use a word for barn, say while translating *Little House on the Prairie,* they have a choice of words that actually mean "granary," "haystack," "grain or olive bin," or "carriage house." When all else fails, they say *cuadra* (as in quadrangle), best translated as "that four-sided thing over there." Another Spanish word that moved into English is *rancho,* the source for "ranch." In Texas, it basically meant what it meant in Spain; a *rancho* is just a camp or, by extension, a casual gathering for some purpose, not a substantial landholding, dwelling, or business.

In the long run the Mexicans made one unique contribution to the American cattle ranches of the High Plains, and that was the art of roping from horseback. This was unknown in Spain, where animals were also captured with a horse, a rope, and a noose. There, the rope with a noose on the end was carried on a long pole, as from a giant fishing rod, and dropped directly on the critter's neck. Throwing the loop, once the skill is acquired, is far superior to running around with a long stick trying to slip a rope over a beast's head. Properly done, using a well-thrown loop to encircle and cinch a pair of the cow's legs, it will immobilize the animal. Even with modern fenced-in stock, the lariat is a useful tool. Once a newborn calf is able to run well, some twenty-four hours after it's born, getting close enough to catch one bare-handed can turn into a marathon. A small lightweight lariat gives a rancher a chance to immobilize the calf from outside its threat zone, before it runs, and then ear-tag and weigh the newborn. Once the calves can run, their speed is considerable and their endurance remarkable.

The Texas system of open-range and year-round foraging didn't stand a chance once it left the semitropics along the Gulf of Mexico. And years before it reached Wyoming and Montana, cattlemen should have understood that it wasn't going to work there either. In the winter of 1871–72, when eastern Montana was still cow-free, actually Indian country, brutal weather killed from 50 percent to 75 percent of the cattle already running in open-range herds in Nebraska and Kansas. A repeat in 1881–82, just as the drives to Montana peaked, killed even more cattle. The only thing unique about the famous winters of 1886–87 and 1887–88, the ones that wiped out the just-established open-range herds from Wyoming up through the Canadian prairie provinces, was its breadth. In 1886–87 animals died from northern Texas all the way to the plains of Alberta, Canada, with losses in herds of up to 90 percent. Under the Tex-Mex system, a cow would find its own feed and water or die. The Miles City, Montana, artist Charles Russell's famous postcard, sent to the absentee owner of the herd—the starving cow, kneeling, ribs showing, variously titled *Waiting for a Chinook* or *The Last of Five Thousand*—was no exaggeration. A Chinook, a warm wind from the west (where the Chinook Indians lived), is still a blessing on the High Plains. It's caused when steady lower-level westerlies blow across the Rocky Mountains and spin down to the plains; the warmth comes from the compression of the air in the downdraft, the same compression that makes a bicycle pump warm when operated.

In general, the settling of the western United States was almost entirely an east-to-west movement. But "Westward the course of Empire takes its way" and Horace Greeley's "Go West, young man," although somewhat accurate, are not the whole story. The California gold rush of the 1850s leapfrogged the Great Plains without settling them. Shortly after arriving in Montana in the 1860s, having come north from California and Oregon following another gold rush, ranchers from Montana's Pacific slope then pushed eastward over the Great Divide and out onto the plains into the traditional Sioux and Blackfoot hunting grounds in central and eastern Montana. By the time of the Battle of the Little Bighorn in 1876, stockmen were settled in the Musselshell River valley. That is so far east into the federally designated Indian hunting area that it is due north of Custer's last mistake.

As soon as the hunting grounds opened up after the war, a number of midwestern and southeastern cattle growers moved to eastern Montana, northern Wyoming, and the Dakotas. Thus two waves of the British Highland ranching style, the first, creeping eastward from the Rockies before 1876 and the second, coming from the settled East, converged. They preceded, surrounded, and quickly outperformed the offshoot of Tex-Mex free-range ranching that bubbled as far north as Canada. In 1879 and 1880, the first years that the mythologized Texas trail drivers were moving cattle down the Powder River to Miles City, they were passing near, and sometimes through, established ranches. To the south of Miles City, right on the trail from Texas, the Evans brothers, both Missouri cattlemen, were in business on the Powder River by 1879, with barbed-wire enclosures that the first drives from Texas had to skirt. The next year, midwesterners founded the Grinnell Ranch on the upper Tongue River, which runs parallel to the Powder and directly into Miles City. By 1885 that organization had spent twenty thousand dollars on ditches and had five thousand acres under irrigation. The Celtic system of raising winter food had arrived as early as the herds of foraging longhorn cattle.

By 1880, even as the era of the trail drives peaked and real ranchers were settling along the trails, Texans had become a small fraction of the plains cattlemen. That year's census for Laramie County, Wyoming, is a good example. It listed 386 men who told the census taker that they were cowboys. Of that crowd, 45 were born in Missouri, another 101 were from the Ohio Valley states, and among the other smatterings of origins, there were only 32 Texans, about 12 percent of the cowpunchers. Actually, Texans weren't cowpunchers. That bit of slang came north from California where the local Hispanic cattlemen did carry blunt lances, as their ancestors had done in Spain, and did poke their cattle to get their attention and get them moving. With Californians following the gold strikes to Montana as early as 1860, it is certain that the word "cowpoke" was spoken in the territory before "cowboy" crossed anyone's lips.

The payoffs were immediate for what the Texans derogatorily called she-ranchers—men who stayed home and cared for their cattle in the winter. It is not clear what the Texas trail riders meant by she-rancher, but ranching with fences and ditches and haystacks would remind a trail rider of nineteenth-century women's work. An irrigated field would resemble a housewife's kitchen garden planted to feed the household, and fences would look like attempts to "civilize" both cowboys and cattle. At the headwaters of the Tongue, now Sheridan, Wyoming, one rancher kept records of the productivity of his winter-fed cows against a neighbor's animals left to forage for themselves Texas-style. In the brutal winters of 1885–86 and 1886–87, his cows outproduced his neighbor's by six to one at spring calving season. And that was in rich grass country where downsweeping Chinook winds from the Bighorn Mountains often cleared and melted the snow. The fencing, irrigating, stock-breeding ranchers, in the literal meaning of the word, were the first *settlers* of Montana. The miners shifted with the strikes, the trail riders came and went, the cowboys moved from ranch to ranch. A few years later, the homesteaders would follow the she-ranchers to the big open country.

Romantic as the movies may make it, the reality is that the trail drivers with their great herds were never pioneers or explorers. They were moving cattle toward civilization—the railroad towns—not away from it. The vast range opened up by the Sioux War of 1876–77 was the attraction, but the big herds came in the early 1880s to meet up with the railroad in Miles City.

The eternal combat between cowboys and Indians is another creation of Hollywood. By the time the trail drives began in earnest, the routes to Wyoming and Montana were essentially conflict-free. Indians along the trail might rustle a few "slow elk," but they were more likely to extort a steer or two in a face-to-face with the trail boss for the promise of safe passage. The only exception to this was along the trail from Texas up through the western edge of Oklahoma Territory and then to points north. Complaints from ranchers on the settled farms in Kansas, who blamed the trailed cattle for outbreaks of disease (the Texas cows had them all, including tick fever, anthrax, and foot-and-mouth), and even armed confrontations between the ranchers and the trail drivers had pushed the cattle herds to a more western route through the deserted plains that surround the Oklahoma-Colorado boundary line. A historian at the University of Wisconsin noted in his 1931 Ph.D. thesis that cowboy-versus-Indian conflict "continued [there] down into the decade of the eighties; for western Oklahoma was about the last place in which the great American outdoor sport of Indian fighting was available."

"Don't Fence Me In" is an iconic cowboy song, but it's from Tin Pan Alley not Miles City or Saskatoon. It is not a tune hummed by cattlemen. To be honest, Miles City, where the author was partly raised, has fallen for the myth. The local historical museum celebrating the countryside's ranching heritage is the Range Riders Museum, which sounds better than the Post-Hole and Ditch-Diggers Museum. The stock business is all about fencing, and as soon as the briefly open range was fenced, the plains opened up to the excellent pedigreed beef animals imported from Britain.

The opening of the High Plains after the Great Sioux War allowed both the open-range cattlemen and the permanent, fenced and plowed, cattle ranches to spread over most of the plain, followed, almost immediately, by the small holders who came and made their Homestead Act claims on what had not yet been settled. The country was occupied, but it was still a work in progress. The heart of the western beef industry would be greatly improved animals, either pure breeds or the fast-growing mixed-breed animals with at least one quality parent. Everything was in place, and the West was waiting for breeding stock that could make the best of the rugged terrain.

Within a decade, three main types of purebred animals were

scattered across the plains from Wyoming to Alberta. First came the red or multicolored English Shorthorns and then the white-faced Herefords and, a few years later, the black Aberdeen-Angus from Scotland. There are other breeds found in significant numbers on the plains today and endless variations of first- and second-generation crosses. A truckload of animals headed for the feedlot may be as parti-colored as a quilt.

But it is the pure herds (always behind fences) that attract the eye. In addition to the striking white-faced, red-haired Herefords and the ebony Aberdeen-Angus, there are Red Angus and a breed called Beefmaster that is about the same shade as Red Angus but comes equipped with horns. Ash-blond Charolais are also present in numbers. In all cases of these purebred herds, the eye is pleased by the congruence of their solid geometry and the similarity of their coloring. They much resemble wild game in those respects. It is not just that cattle have replaced the buffalo physically, but when they are in handsome purebred herds, they come close to replacing the buffalo aesthetically. The crossbreds seem much more like intruders. Wild animals are naturally all of a size and a color, save the occasional oddity like a white buffalo. The storied Longhorns were variously mottled, brindled, dappled, and just plain nondescript. A good herd of purebred beef critters is all of a piece, as much so as a band of antelope or mule deer. One is reminded of the poet William Blake's tiger: "What immortal hand or eye / Shaped thy fearful symmetry . . ."

We would likely answer "random genetic variation and natural selection" produced the great striped cat unless we were strict Creationists. It is different with cattle. We know who did it. It was three Scotsmen in particular, and their fellow likeminded farmers, who made the Aberdeen-Angus the excellent, symmetrical, predictable, and recognizable animal it is today. They made the breed uniform.

→ 4 ←

In Scotland, a Great Breed Is Born

WHEN cattlemen speak well of a good herd of beef cattle, of any breed, uniformity is always remarked on. Animals of the same age, before life's various vicissitudes and rewards remodel their bodies, should be practically the same size and weight. A Western rancher, and especially a British farmer, may have cattle with such distinct qualities uniformly found through his herd that the animals look like clones. Western ranches tend to have more variety, partly by deliberate choice and partly because there are more important qualities for a range animal than being cookie-cutter identical to its herd-mates. Like any trait, uniformity can be pushed too far. But in good herds, the distinctions among animals are subtle, and to the still-uneducated eye, they may look exactly the same. A breeding-stock ranch's Aberdeen-Angus heifers, in particular, and bull calves, to almost the same extent, make up herds of very similar animals. (Bull calves and heifers are separated early in life to prevent the accidental inbreeding of animals that reach sexual maturity ahead of schedule. By the age of animal adolescence, they are rigorously kept apart.) The heifer and bull herds are as tolerant and even desirous of contact as all cattle, and when a herd moves, it coalesces into a smooth and solid and singular black entity. At a trot they have a smooth gait, and the topline of the massed herd does not undulate as it draws a black streak across the ranchland. By the way, just calling them calves gives the wrong mental picture of a herd of bull calves. By the time they are ready for sale, bull calves weigh a half-ton even when lightly fed with supplements and otherwise required to forage across the plains. They are no longer cute calves, but the language of stockmen lacks a specific word for this sexually mature but still-filling-out beast. And as far as sexual impulses are concerned, bull calves would put a junior high school band to shame.

The same morphological consistency is found in other breeds, including the color variant of purebred and registered Red Angus in the United States. (The Aberdeen-Angus Society of Great Britain does not discriminate, so there is no need to form a separate registry there.) Herefords, both horned and polled, and several of the minor breeds approach and occasionally in some herds surpass the conformity of an average herd of Aberdeen-Angus. There are two basic reasons for seeking uniformity, one for the registered breeding-stock rancher, another for the commercial raising-for-meat operator.

In the breeding-stock herd, a pleasing uniformity suggests (although it does not guarantee) that the purchaser is getting a fair sample of good genes no matter which animal he chooses. This consistency in form also indicates that the breeder has done his homework. There are times when purchasers do want an animal that is exceptional in some direction, say length or total weight, if they need those particular genes to improve their own stock. This can be carried to extremes, as we will see later when we look at some twentieth-century fads in Aberdeen-

Angus breeding stock. And an animal that is exceptional in some visible way may also be unusual in a hidden quality, particularly behavior. There is no question that temperament in cattle is an inheritable trait.

For the commercial stockman, the one who is turning out all that Certified Angus Beef in North America or product for the Aberdeen Angus Beef Club of the British Isles (meat brand-labeled and sold in supermarkets in each country), consistency means he will get the growth he needs without turning out the occasional little fellow or, and it is also a penalty in the North American marketplace, the too-large steer that doesn't lend itself to neat shipping-box packaging for supermarkets or cookie-cutter steaks for the portion-controlled restaurant industry. A good deal of feedlot-bound steers and heifers fend for themselves on leased, distant pastures every summer, not seen by the rancher again until the fall roundup. Under those conditions, in a good grass year, some animals can get too big without being culled and sent to market in a timely, and more profitable, manner.

When Aberdeen-Angus men talk about the perfection of the breed, which took place roughly between 1820 and 1880, they are speaking of the search for conformity by the acknowledged "fathers" of the modern cow. Indeed, the very first of the master breeders demanded conformity in the animal's most obvious character, its ebony coat of sleek hair. (Its polled head is also obvious, but many types of old Scottish cattle were both hornless and multicolored. A recessive gene for a red coat is still present in Britain, and in America herds of homologous Red Angus are much prized by their breeders.)

Hugh Watson of Keillor (1789–1861), a parish in Angus, is always acknowledged as the first great improver of the breed. He insisted on the pure-black animal. When he began to farm at the age of nineteen, he is said to have begun with a bull and six cows from his father's herd, animals that his daughter recalled him saying were chosen not only as the best but the blackest. Blackness was common in some herds, but the original Angus or Aberdeen polled animals also came in red, a sort of brindle (a variegated coat of tan and black with a touch of reddish blond), what one writer described as

Hugh Watson

"mousey grey," and frequently in mottled black-and-white. The-black-and-white animals were particularly prominent in far northern Scotland, a destination for many years of invading settlers from the southern shore of the Baltic, from Friesland on the west to Denmark on the east. It is highly likely that anyone coming from that area to the Celtic shore would bring along some splotchy black-and-white cows. The common dairy cow all over the world, the archetypal cow, is the well-horned Holstein-Friesian that has lent its black-and-white coloration to countless mailboxes and Gateway Computer packages, and without doubt it is related to the mottled animals from Orkney and the Shetlands and the northern coast from Inverness to Aberdeen. As you will see later, the kinship of polled Aberdeen

Angus and the horned Holstein-Friesians is historically close, and it may have come uncomfortably close toward the end of the century just past.

Other than an insistence on blackness, little is known of the principles that Hugh Watson applied when selecting animals and mating bulls with cows. He registered very few animals with the secretary of the Royal Highland Breed and Agricultural Society in Edinburgh. When the first *Polled Cattle Herd Book* was published in 1866, it picked up polled registrations from the Edinburgh records, and his cow Old Grannie and his bull Old Jock were the first animals of either sex entered on the new register. A breed registry, like a farmer's herd book, will include information on the parentage of each animal, at least back to the grandparents. A bull registered tomorrow with the Aberdeen-Angus Society of Great Britain or the American Angus Association can be traced back to its most distant registered ancestors. What we do know of Watson's breeding is that after purchasing another bull and a few heifers in 1810, only the second year of his herd, he never again brought in an animal from another breeder. With a herd that never numbered more than a few score, he somehow managed to avoid the usual ill effects of inbreeding (loss of vigor and growth, as well as genetic defects). That is as much a tribute to the great variety of vigorous ancestors in the existing polled cattle, which had been rather casually mated for centuries, as it is to Watson's astuteness.

The insistence on pure black may have accidentally increased the gene pool in Watson's limited sample of the world of Angus cattle. Authors familiar with Scotch cows before the breeds were regularized and registered all mention that there were two distinct types of polled Scottish cows that appeared in Angus and also in Aberdeen. One was a diminutive animal usually kept by sharecroppers (crofters) as a milk cow, a small and thrifty beast that could survive on poor pasture, and little of it. These small polled cows were invariably and entirely black. The other polled cattle, used for beef and very often as milk cows, were much larger animals meant for better lowland pastures. The big ones were the type of Aberdeen and Angus cows that came in several colors as well as basic black. Even the best of the large almost-all-black cows carried white on the centerline of their underbody, including, as appropriate to their sex, the udder or the scrotum.

Watson's first females were as black as possible but not entirely so. His first two bulls are thought to have been totally black. It seems entirely possible that the all-black bulls, coming from a stock otherwise parti-colored if not entirely without blackness, would have included some of the diminutive pure-black Highland hummlie in their ancestry. As would be discovered much later when small black hummlies were bred deliberately, if Aberdeen and Angus bulls are put to full-size cows of another breed, the offspring can be vigorous and even larger than the dam's type. In any case, it is a historical fact that the perfected Aberdeen-Angus cattle carried within their DNA the possibility of reverting to the truly small animals that nineteenth-century breeders dismissed as a poor folk's poor excuse for a cow.

When the first polled registry was under preparation toward the end of Watson's life, he took offense at some real or imagined slights and refused to give the editor a single bit of information beyond what could be gleaned from Watson's scattered registrations with the Royal Highland Society. For his own use, Watson would have kept a complete record of every breeding animal and its offspring. His widow went a step further: On Watson's death she burned the farm's herd books.

What information survived was as much as his son could recall of the exact lineage of the Keillor herd.

The level of inbreeding must have been extreme. Not only did Watson stop purchasing animals from other farmers, but some of his own cows came to dominate within the herd. Old Grannie produced twenty-five calves herself, all of them retained in the herd, where some of them founded notable lines, or families, within the breed. The sole purchased bull that Watson brought into the herd in 1810, Tarnity Jock, was the male ancestor of most of Watson's best animals. As will happen, whether it is a trait of size, rate of growth, or behavior, a sire may influence daughters more than sons, or vice versa. Tarnity Jock's genes produced several of the most famous female lines in Aberdeen-Angus history, names known to every breeder: Erica, Pride of Aberdeen, Beauty of Tillyfour (bred at Keillor by Watson and sold to the Tillyfour herd), Jilt, and Miss Watson. To this day, at any auction of registered Aberdeen-Angus, a buyer won't have to turn many pages in the catalog before seeing males and females with names beginning with E (for Erica) or J (for Jilt) and animals that are Pride of this and Pride of that. On both sides of the Atlantic, Ericas appear, usually with the name of the farm or ranch included. The British style is "Erica of Suchandso." Americans tend to put their ranch's name prominently in front, often abbreviated: "FAR Erica," with an attendant registration number, means a cow from the Erica family raised on the Felton Angus Ranch.

In some respects this is vanity, for an animal today, separated by some one hundred generations from these venerable ancestors, has an infinitesimal amount of Erica germ-line blood or Pride of Aberdeen genes. However, FAR Erica would have mitochondrial DNA (mtDNA) that closely resembled the original Erica's mtDNA. This DNA—quite separate from the DNA inside the nucleus—and everything else in the outer shell of the cow's cells is passed on directly to both sons and daughters by the mother. The female's egg capsule holds tiny organelles to do most of the energy conversion for the cell's growth and maintenance. This mitochondrial "factory" makes the various hormones and other chemicals that pass out of the cell and are distributed throughout the organism. DNA, which gets all the credit for being the controller of the biological universe, is just that: DNA is management; the mitochondria is labor. Only inside the nucleus does one find the mixtures of male and female parental chromosomal DNA that make up the various and variable individual offspring. The only thing that changes mtDNA as it is handed down from the mother to all of her offspring is random mutation over time. A supermarket slice of Certified Angus Beef may have mitochondria scarcely different from Old Grannie's or barely distinguishable from the original Erica's. It is this consistency over eons that allows genetic anthropologists to talk about an African Eve but never about any single ancestral Adam.

Although the early Aberdeen-Angus breeders were totally unaware that a major component of every animal they bred was the dam's mtDNA, a concept centuries in the future, they instinctively understood the remarkable effect that a good cow had on her calves, male and female alike. Of all breeds being regularized in the nineteenth century, only the Aberdeen-Angus strains were routinely identified not by the sire of the herd but by the dam. To be sure, a great bull was immediately effective in improving the quality of a herd, as in a single year he could breed dozens of cows. But Scots then, and some Scots today, will advertise their breeding-stock herd as being, for example, of the Erica family or the Jilt family in those rare instances where the farmer adhered to the standard of the breed and did not move in some trendy direction. This emphasis on the female line and, in particular, on keeping the offspring of excellent females within the herd has ramifications that still appear in the breed.

When the mtDNA of a large sample of Aberdeen-Angus cattle is compared to other breeds, its mtDNA looks quite different. It has more unique sequences of those boring acid molecules than do other breeds, and Aberdeen-Angus share these unique sequences throughout the breed. Taken as a whole, Aberdeen-Angus mtDNA clusters together exclusively, distinguishable from the mtDNA sequences of other modern breeds. The only breed that approaches it in being different from the crowd is the almost extinct Iceland cow.

When the mtDNA is highly similar within its own group, this suggests what is called a genetic bottleneck, meaning that a very few number of female ancestors made it from one center of population to the new one. It is this bottleneck that worries biologists when members of a species approach extinction and the only remaining individuals are ones that migrated to an out-of-the-way place and survived undisturbed. These remnant populations are almost always drawn from a very small number of "founders." A bottleneck makes particular sense for the Icelandic cattle; it's a long way to row a cow from Denmark to Iceland, and it was done as few times as possible. In the case of the Aberdeen-Angus cattle, not only were they originally drawn from small and localized populations centuries before now, but the emphasis on the female lines during the nineteenth century squeezed the neck even tighter as the offspring of just a few cows, generation after generation, were preferred and came to dominate the breed.

The breeders' interest in female lines has carried into the twenty-first century, except that now it is calculated mathematically. When Aberdeen-Angus breeding stock is advertised to-day, a series of complicated estimates of the animal's genetic potential, compared to a theoretical average, is always included. Shortened as EPD in the United States, it is the "expected progeny difference" from what a theoretical "average" parent would contribute. For instance, a bull from a line that turns out smaller-than-average calves will have a negative EPD for birth weight; an EPD of –3.0 means his calves will weigh that many pounds less than the breed's standard of comparison. The other most commonly used EPDs are for milk production, weaning weight (6 months), and yearling weight. Other breeds do the same (copying the American Angus Association's innovation), but only with Aberdeen-Angus does the buyer get a genetic EPD for the dam—for the female line. The Hereford breeders, particularly in Canada, are considering paying attention to female EPDs, but this comes only decades after Angus breeders in the United States created the system. This emphasis on the female families goes back beyond recorded history with Aberdeen-Angus, although quantifying it is a late-twentieth-century innovation.

Calculating EPDs can turn up an unexpected genetic potential. Initially, only heifer calves were assigned an EPD for milk, on the assumption that it was a maternal, female-only, characteristic. But when the predicted outcomes weren't consistent, a little research indicated that the bull had as much to do with milk production as did the cow. This shouldn't have come as such a surprise. The importance of the sire was long appreciated by dairymen, but cattlemen pay little attention to the dairy business.

If Watson was the first notable breeder, he was only part of a much larger and dispersed group of farmers in northeastern Scotland who kept black cows. A famous example is the herd at

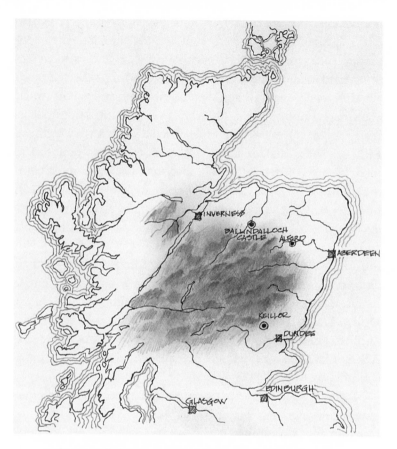

**A map showing towns in the shire of Aberdeen
and the former district of Angus**

Ballindalloch Castle in Aberdeen, which had been all black
(with the perfectly normal bit of white here and there) and all
polled as long as anyone knew, back even to the late Middle
Ages when the castle was little more than a "keep," a strong-
hold for a highland chieftain and a few loyal followers.
Other, less prominent farmers also had predominantly black
animals, and they would be the source for some of the best

genes (and mtDNA) ever worked into the more famous herds.

The second of the triumvirate of founding fathers was a man who knew practically every farmer in Scotland with black cattle for sale. William M'Combie was born into a family that made its living from buying, driving to market, and selling "lean cattle," and he knew from experience that the cattle we now call

William M'Combie

Aberdeen-Angus were the best value he could find in Scotland. His father, Charles M'Combie, the largest dealer and drover in all of Scotland, did well enough to purchase a large farm at Tillyfour in Aberdeen, some sixty miles north of Watson's Keillor farm in Angus. Droving was a rough business, and Charles M'Combie sent William off to college in 1819 to prepare for a gentler life. His older brother also went to college and became the Presbyterian minister of Alford, the parish church of Tillyfour. It is understandable that the senior M'Combie would encourage both his sons to seek a gentler lifestyle. One indication of the very real dangers in buying and selling cattle for cash and driving them through the wildest countryside was that the registered drovers and dealers were exempted from the general disarming of the Scots, in 1716 and in the second disarming act of 1748, after the uprising under Bonnie Prince Charlie: For most of a century, soldiers, cattle drovers, and, illegally, highwaymen were the only Scots carrying guns on the open road.

Two years at Edinburgh was enough for both William and the university, and he joined his father in the cattle business in 1821. In 1824 he leased Tillyfour from his father (by the then-in-effect rules of primogeniture, William's older brother would inherit the farm and be McCombie's landlord for the rest of William's life) and began raising cattle—mostly, but not entirely, Aberdeens. He chose them for strictly utilitarian purposes and was not so particular about pure blackness as other breeders. A bit of white, particularly on the underline, he thought was a sign of potential excellence. He favored the polled animals for their ease of handling and their hardiness and, in 1830, established the herd of purebred Aberdeen cattle that became the standard of excellence in the mid–nineteenth century. He and his fellow breeders found a demanding local market for black hummlies, but the great incentive for the breeders was the fundamental change in the scope of their market. Advances in transportation allowed Scotland to progress from an exporter of lean cattle at minimal price to the home of a thriving industry in fattened animals sold at a premium.

When Scots started feeding their own cattle for market, rather than selling them lean and driving them south to fatten, they perfected the techniques of agriculture—fertilizing, crop selection, drainage of fields—just in time to take advantage of the changes in transportation effected by the steam engine. By the 1840s small local freight railroads connected with new dockyards serving the new steamship carriers (most built to move coal, with cattle as an afterthought). By the 1850s Scotland's railroads began to connect directly with the English railroads, eliminating the need for cattle boats, always subject to the vagaries of weather in the English Channel and the Irish Sea.

Gone was the droving business entirely, and fattened cattle went to market without losing weight as they did on the old, rigorous drives. Changes in agricultural science were as striking as the machines created by applying physics and chemistry to manufacturing. It was as much an age of agricultural revolution as it was industrial. As already noted, turnips had become the plant of choice for winter feeding and fattening. Turnip yields improved immensely when additional phosphorus was added to Scotch farmlands, either from domestic bonemeal (perhaps the earliest example of recycling) or from the powerful nitrogen and phosphorus found in South American guano, by then available cheaply because of the great improvements in transportation.

If Watson began the purebred business, it was M'Combie who first recognized a very real threat to the Aberdeen and Angus breeds, a threat caused by the very success of the breed and the almost simultaneous opening of the English market to fat Scottish stock. The native black animals made such marvelous crosses with other breeds that farmers were tempted to dilute the purity of their black cattle as they turned out crossbreds, males for the butcher and heifers for breeding. M'Combie, a founder of the Polled Cattle Society (prelude to the Aberdeen-Angus Society of Great Britain) understood in those days when the word "gene" didn't exist that the success of crossbreeding for hybrid vigor was absolutely dependent on the purity of at least one of the parents, the bull or the cow. It didn't matter greatly which was the purebred.

The most successful raisers of beef cattle for the market (and some of the most successful animals at the fat stock shows that had become popular) depended on crosses of Aberdeen-Angus cattle with British Shorthorns. There was a distinct difference in the longevity, the conservation over generations, of the hybrid vigor, depending on whether the sire was a Shorthorn put on

Aberdeen-Angus cows or a good black bull put on Shorthorn females. It was this difference that increased the threat to the purebred herds. In the first generation of crosses, the offspring of either kind of mating were equally vigorous. Making steers of the males and keeping the hybrid heifers for breeding worked for several generations with a Shorthorn bull put on hybrid Shorthorn cross Aberdeen-Angus females. A farmer could keep as many females as he wanted from his own herd, use the same bull or another black sleekie, and continue to turn out good males for the meat market and females for his breeding stock for several generations. There was every incentive for the farmer to increase his herd by keeping and breeding his crossbred females. Eventually, vigor was lost and parti-colored animals, even some with horns, came out of the mixed-blood mothers. That made it necessary to find purebred Aberdeen-Angus cows and bulls and start over again. The same eventual decline in crossbred vigor happened with Aberdeen-Angus bulls put on Shorthorn cows, but the lack of vigor could begin as early as the second generation. The very first calf out of a first-generation hybrid—that is, the result of a male Aberdeen-Angus on a first cross (F1) hybrid female—could be a merely satisfactory animal, lacking the exceptional hybrid vigor seen in the first generation. This remains true and mysterious, but it is one more piece of evidence pointing to the importance of maternal lines in breeding.

The continued good performance of the generations of mixed-breed cows put to a Shorthorn bull through several crosses threatened the Aberdeen-Angus. By the time farmers began to realize their calves were regressing, they might have no pure black cattle at all. In these cases, the founding mothers had died, and the stockman had no purebreds to work with in improving his livestock. It was M'Combie and a handful of other breeders who kept with pure stock who made it possible, when reason prevailed, to find good and pure mothers and excellent purebred bulls, whether the purchaser wanted to raise commercial all-black cows or hybrids or wanted to get into the purebred Aberdeen-Angus business.

M'Combie himself raised a few F1 hybrids for commercial meat and occasionally even showed hybrids in the bullock (i.e., neutered male) class at the important shows in Britain. He was a man of considerable modesty, given that he was acknowledged in his lifetime as the leading breeder, even to the point of providing the foundation stock for Queen Victoria's Aberdeen-Angus herd. He liked to say that he was more of an expert at fattening cattle than at improving them. "I can hardly speak with the same authority," he wrote in his autobiographical *Cattle and Cattle-Breeders,* "as a breeder, generally, that I can as a feeder."

Indeed, much of the section in the book where he describes his breeding style is not about his animals as much as an encomium to breeders he most admired. Although M'Combie was easily the greatest showman of fat cattle in his era, he particularly respected stockmen who did not force-feed their cattle for fairs or who simply did not enter the great fame- and money-making shows. "Mr. Mustard," he wrote in a passage that typifies his attitude, "[of] Louchland, is a very old breeder, and I believe no purer stock exists in Forfarshire. Mr. Mustard never forces his stock for the show-yard, and seldom sends any except to the county show, where they are always winners." He also commended a Lord Airlie, who "wisely sets himself against forcing for the show-yard," as does the owner of the Auchlossan herd who "hates the forcing system, he has taken many prizes at the Royal Northern Agricultural Society's shows . . . with cattle in their natural condition."

A visitor to a Scottish herd (and some American ranches) today will find the same distinction. There are British breeders whose success in producing a big animal is partly owing to overfeeding (forcing), and there are wiser men whose large cattle are raised on the historic diet of pasture forage from spring to fall and then dried hay and turnips in the winter, complemented with small amounts of grain supplement. A good Aberdeen-Angus strain has a remarkable ability to fatten on rich pasture alone. On a farm in Aberdeen, visitors, including the author, noted at least one heifer that needed to be moved to a previously grazed field on the farm: She was in danger of getting overfat on nothing but the abundant clover and grass in her paddock. Fatness can interfere with calving and with milk production (fat accumulates in the udder and displaces milk-producing glands). That is a pleasant problem to have to deal with—a cow that does too well without any supplemental feeding.

Whereas Hugh Watson's style of breeding at Keillor was deliberately a mystery, M'Combie, a founder of the Polled Cattle Society, registered his animals and shared his herd book with colleagues and prospective customers. He founded his purebred herd in 1840 and made the greatest contribution to it and his greatest improvement to the breed in 1844, purchasing a heifer named Queen Mother from a man whose name would be otherwise extinguished in history: "It is to Mr. Fullerton," he wrote, "that I owe my success as a breeder. I shall always look up to him as the founder of my stock." Again, as with so many events in the history of the breed, it is a maternal line that matters.

M'Combie's greatest success, founding the first of several "Pride" families with descendants of Mr. Fullerton's Queen Mother, came about by his constant infusion of new blood into his herd, quite the contrary of Watson's severe inbreeding. The

M'Combie method is the one followed ever since by the successful breeders of registered Aberdeen-Angus cattle. "The breeder ought to be always buying and selling and incorporating different *strains* together. [That is, strains of Aberdeen-Angus, not different breeds.] There will be many blanks," M'Combie wrote, "but there will be a prize; and when you hit, and when the incorporation proves a lasting benefit and is stamped on the original herd, it is a great prize you have won." He used the same verb "hit" that one would use for winning a lottery; he acknowledged that no matter how much he studied pedigree, it was something of a gamble. (M'Combie himself was a noted gambler, the host of card parties and associated revelry that would go on for days.)

And M'Combie knew well the dangers of pushing too far in a single direction and neglecting the breed's natural good points; indeed he confessed to doing it. Reflecting on his purchase of a bull for its conformation alone, without considering the animal's behavior and the behavior of its parents, he realized he had purchased a handsome animal with an ugly and very heritable temperament: He ruefully wrote that "Docility in temper in male and female is indispensable. Inexpressible mischief may be done by the introduction of wild blood into the herd, for it is sure to be inherited. I have suffered seriously by this error." This is, as they say in Montana, producing "ringy" stock. Weeding it out of a herd is expensive, since otherwise more profitable, higher-priced bull calves are castrated and sold for beef, and suspect heifers are either sent off to the packing plant or sold off without pedigree as "commercial" stock.

M'Combie maintained some sense of humor about his own lapses in judgment. Humor is a natural Scottish attribute. The "dour Scot" of literature is a somewhat, though not impossibly, difficult creature to find in the flesh. M'Combie had won the most prestigious of Scottish shows, the Royal Northern Agricultural Society's fair, for two consecutive years with two different bulls. A third victory with a new bull would retire that cup, and it would reside with him in perpetuity at Tillyfour. When the moment came, M'Combie was beaten out by a bull he had bred, not particularly admired, and therefore culled. He sold the calf to the stockman at Castle Fraser, only to see the animal come back to deny him the only prize he still coveted. "It [the cup] was my last asking, but was dashed from my lips, and went for the time to Castle Fraser, instead of going to Tillyfour forever." Not only could M'Combie confess his own mistakes, but he openly admired other men's better judgment. He related the case of a neighbor identified only as Mr. Brown, whose "skill was tested as to the purchase and sale

Sir George MacPherson-Grant

of [his champion bull] 'Windsor.' He bought him from me as a calf in poor condition for under 40£, and sold him to Lord Southesk for 200 guineas [210£]."

The Lord Southesk (whose estate at Kinnaird Castle on the South Esk River would eventually become a rather quirky hotel) might well have gone down in history as one of the great improvers of the breed had it not been for an outbreak of rinderpest that completely wiped out his herd. Southesk certainly bred one of the most famous dams, if not the greatest of all time, and sold her to Ballindalloch Castle just a few years before the plague struck his renowned herd. This was Erica, sold by Southesk to the baronet of Ballindalloch, Sir George MacPherson-Grant, in 1861. She was the cow that elevated the Ballindalloch herd from respectability to international fame. "Volumes could be written," a historian of the breed wrote in the society's journal, "about Erica whose prolific family has been so literally worked into the whole fabric of the breed that a history of her descendants would be almost equivalent to a history of every [British] herd." Putting it another way, the current secretary of the Aberdeen-Angus Society of Great Britain remarked that a computer printout of Erica's descendants and their pedigrees from her down the generations to her most recent relatives would be several miles long.

They still raise Aberdeen-Angus at Ballindalloch. Since the turn of this century, there's a new herdsman and a new general manager with a fresh commitment to excellence. It is certainly the oldest herd in the world, not just continuous since registrations began in the nineteenth century but going back beyond memory. William M'Combie thought it was the oldest herd of polled animals in the north, and wrote that "it has been the talk of the country since my earliest recollection and was then superior to all other stock." The new herdsman at Ballindalloch, Colin Sutherland, could not make so extravagant a claim today, but he is dead set on returning the estate's animals to the top rung of herds. Unlike many of his contemporaries, Sutherland has a good basic stock of the old-fashioned large animals to work with as a foundation, the kind of beast once again eagerly desired by breeders and commercial herdsmen.

There is a painting hanging in Ballindalloch Castle of a

Ballindalloch Castle

group of cows, a painting dating from more than a century ago. It would be Sutherland's triumph if he could breed as fine a group of cows in this century. That is not to say that the Victorian era was the apogee of Aberdeen Angus cattle. There were certainly "perfect" individuals and very good herds in the 1800s, but the breed as a whole was less consistent than it would become. Raising cattle is not unlike other manufacturing businesses. All proprietors want to produce a good product, but it has to have a market. And to fulfill the market's demand, both cattlemen and factory managers must produce a consistent product in a predictable manner with a minimum of waste. That is why the "perfection of the breed" is a constant and continuing process.

→ 5 ←

Cow Wars of the Nineteenth Century: How the Aberdeen-Angus Conquered the Cattle Shows and Defeated Its Only Serious Rival

ALTHOUGH everything about a High Plains ranch seems not just an ocean away but a world away from the small, intimate cow farms of Scotland, there are hundreds of North American Aberdeen-Angus that get as much handling and are raised in as close contact with humans as any of the beasts kept by the founders of the breed in Scotland. These are the animals raised by children and adolescents who belong to either 4-H or Future Farmers of America. Youths exhibit their animals in competitions beginning in their local chapters right on up to statewide contests. Most are ranch kids, but anyone with enough money to buy a calf and enough backyard to pen it can be in the game. There are all kinds of contests for all kinds of animals, including for best cow and calf, or best bull, cow, and calf. But the largest and possibly most lucrative class is for neutered beef animals. Grand champion steers are always purchased for considerably more than their actual value, as steak houses and specialty butchers vie for the publicity that comes from laying claim to the best beef animal in the county or the state. Farm kids understand from day one that their steer is just dead meat walking, but it is easy to turn one into a pet. Animals that are shown are handled early and often, broken to the lead, scrubbed, trimmed, and brushed, nursed in sickness and pampered in health.

Like beauty contests of any sort, cattle judging has fads or trends. People who raise registered breeding stock are ambivalent about cattle shows. Championships can mean money, but they can also lead to a distortion of the breed by a collective faddishness: It has been so since the very beginning.

The first great nationwide cattle show in Great Britain was organized at the end of the eighteenth century by the Smithfield Club, an assortment of baronets, earls, lords, and dukes. That peculiar British phrase "landed gentry" was applicable to members of the Smithfield, as they had estates as well as farm animals. Their ideal in a beast intended for the table was hugeness and the attendant fattiness. Winning prizes at shows, like all empty but seductive activities, became something to brag about at one's dining table, preferably while serving the choicer parts of the champion animal to envious neighbors.

By the time of only the second Smithfield show in 1800, criticism of the landed gentry's criteria had begun. Writing with an attitude of "no small degree of alarm," an anonymous satirist in *Commercial and Agricultural Magazine* called the event a "*raree* show," populated by "overgrown cattle." It was something of a class war, the magazine implied, headlining this criticism of Smithfield as "The Humble Petition of 500,000 Frugally Disposed Housekeepers, Resident in the United Kingdoms of England, Scotland and Ireland."

Similar criticisms echoed down through the whole of the

nineteenth century. A writer in 1849 described that year's Smithfield animals as "mere fat, unwieldy, imbecile brutes." A journalist in one of Fleet Street's livelier broadsheets said of the same show that "Dukes and Earls" were presenting nothing but collected "animal luxuries." They could afford anything, the writer sneered, "whether it is the purchase of a Titian, or the production of a prize bullock." He thought it a victimless crime against good judgment, and useless for real farmers who would of necessity "confine themselves to what is moderate and profitable." Others were not so sure it was a harmless vanity fair. While the aristocrats liked to claim that they were only demonstrating the speed at which animals from their herds could grow and fatten, and while they believed the same growth curve would apply to moderately fed animals, others feared that the obscenely obese beasts would become so popular as to infect and malform the commercial herds with their rampant genetic tendency to excessive size and weight.

Responding to repeated criticism of this tallowmania, the Royal Agricultural Society at its first exhibition in 1839 ordered the judges to look at the breeding potential of the stock, not its gross weight. But within a few years the RAS was judging as if it was a clone of the Smithfield Club. This might have been predicted, for in the first paper ever read to the Society, the president, Philip Pusey, made the Smithfield aristocrats' philosophy his own: "The power of reaching that excessive size [too fat for the real world of markets, as he had earlier acknowledged] is the only test by which the capacity for acquiring useful condition, at the cheapest expense and at the earliest age, can be tried under . . . competition." Serious breeders for the markets, however, paid little attention to the grotesquely fat winners at the shows and even less to putting their own show animals on gargantuan diets.

The reputation of some of the early Shorthorn herds was founded on Smithfield-sized animals, but there was a series of built-in checks against the genetic pollution of England's herds that were either ignored or unimagined amid the rage for fattened kine. Early obesity has near-calamitous results in both bulls and heifers. With the females, it is a generalized difficulty with getting and staying pregnant. Once pregnant, females experience a diminution of milk production because of fatty compression of the milk glands. For bulls, early obesity produces smaller testicles and constricted sperm ducts, and reconditioning a bull after his showtime is over will never replace his capacity for producing sperm in the quantity and quality of a healthy trim beast. By their own excesses, the worst animals had no capacity to breed vigorously.

Not that the Aberdeen-Angus breeders shunned the fat shows. The famous M'Combie of Tillyford took prizes at Smithfield in the bullock, i.e., steer, classes. But he did fat stock only with castrated animals, while at the same time maintaining a healthier breeding herd. In 1856, in his first great triumph abroad, he swept every class at the Paris Exposition, taking prizes for beef cattle in the bull, heifer, steer, and breeding-family classes, out-pointing cattle from all over Europe. When that year's British contingent of Shorthorns arrived in Paris, the judges excluded them from competition as "being too fat . . . not likely to breed." The following year M'Combie took the fat stock prize at Smithfield with a black steer that weighed 2,744 pounds (1,245 kilograms). He was not above producing a morbidly fat animal, but his style was still different from that of his Shorthorn and crossbred competitors. Whereas they gloried in creating animals that grew fat within two years (for the growth potential), M'Combie was a patient sort: His 1857 winner was nearly five years old. Mature animals, he said, were what the butchers wanted.

"I have always been partial to aged cattle," he told the readers of his autobiography, "and if you want a quick clearance, age is of great consequence. The great retail London butchers are not partial to 'the two teeths' as they call them, and I have seen them on the great Christmas day [sale] examining the mouths of cattle before they would buy them. They [young cattle] die badly as to internal fat, and are generally light on the fore-rib." The fore-rib is a bit of a mysterious word; it actually means the rear end of the ribs proper, but it is for(e)ward of the loin.

The expression "two teeths" is clearer. Calves, like other mammalian babies, start out with milk teeth and gradually acquire permanent ones. A calf's first milk set of incisors, all on the lower jaw, is a semicircle of eight skinny pegs. Cows have no incisors on the upper, just a flat bony surface called the incisive bone that the lower incisors use to cut against, something like the flat brass plate on anvil-style pruning shears. Calves get their first pair of their large permanent incisors by two years of age, hence the "two teeths" argot for animals only two years of age. At two and a half years, the second pair sprouts, so that as they near three years of age, they have four permanent incisors. All eight are in place by the end of the fourth year. At stock shows the age of the animal is important, for it does show growth rate, and veterinarians are called in to attest to the age of the cattle. This requires getting their thumbs in through the lips and rubbing along the animal's lower teeth, counting the number of erupted permanent incisors. Some cattle don't mind, none seem to enjoy it, and a few will pin one or more of the veterinarian's distal phalanges between their incisors and incisive plate, a reason cattle-show vets occasionally sport blackened nails.

Besides morbid obesity, the overemphasis on size and fat in bulls and heifers almost required inbreeding, with the usual consequences of reduced vigor and increased genetic problems after several generations. One of the most famous Shorthorn lines, developed by the Collins brothers of Bakewell, England, declined in herds where strict inbreeding, intended to preserve the purity of that particular strain and to make fat cattle, saw a swift and inevitable regression to the unperfected state.

The general public's disgust with overfat animals had occasional influence at Smithfield and the shows sponsored by the Royal Agricultural Society. Every few years correspondents would note that the animals were not so grossly overweight as in the past. That this improvement was repeatedly remarked upon through the decades suggests that the plague of lard at Smithfield was difficult to conquer. In the long run the herdsmen succeeded in preserving useful strains by entering regional fairs, where the organizers were not earls and dukes but small landholders and professional husbanders working for profit-making farms.

Toward the end of the century, organizations like the Yorkshire Agricultural Society were setting standards that included ability to breed and laying out rules describing anatomical proportions appropriate to animals in a working herd. The Polled Cattle Society, a group of (mostly) Aberdeen and Angus farmers and herdsmen, incorporated itself in 1879. It held its own shows, rewarding normally conformed cattle while lobbying the all-breed shows for more reasonable standards. Britain would remain a class society, but the professional herdsmen would regain control of the fairs and breed competitions.

It is also possible that what finally turned the judges against the lard of the showrings was a decline during the second half of the century in the industrial value of tallow. More and more vegetable oils found their way into the food chain, and the triumph of petroleum products in the industrial world added to the loss of premium value in animal fat. No longer would the military preserve its weapons in cattle grease or the soap maker rely exclusively on animal fats. When one left the home lights burning, it would be a gas lamp or a paraffin candle, not a tallow taper. (The rich, of course, had no use for tallow candles— for them the sweeter-smelling beeswax candles and sperm whale oil in their lamps for bright lighting.)

From the earliest days of fat-stock shows, breeders of Aberdeen and Angus black hummlies managed to capture a share of the prizes and the acclaim. Hugh Watson of Keillor, the first great breeder of the Angus strain, took the occasional ox to shows. One of his steers, a neutered son of his best cow, Old Grannie, took first prize at the 1843 Highland Show, held that year in Dundee. The animal walked the thirteen miles from Keillor to Dundee the day before the judging, a feat the obese English Shorthorn champions could never emulate. Indeed, many a Smithfield entrant was brought into the showring in a low-slung wagon, something like the flatbed trailers used today to move heavy equipment. Later that year, Watson shipped the ox to the Royal Ireland show in Belfast (Ireland was then united). There, Victoria's prince consort, Albert, took a fancy to the animal and bought him on the spot. The prince consort kept the beast at Windsor Castle for a year and then entered him in the aristocratic Smithfield Club's Christmas show in 1844, not as a competitor but as "extra stock" on exhibit for general admiration. (It would have been unseemly for the prince consort to enter for the gold medal.)

When the show was over, Prince Albert sold his ox to a London butcher. Before going home, Albert stopped by to give his ox a pat or two when, it is related, "the gentle animal tried to lick his hand; and the Queen . . . immediately gave orders that the animal was to be repurchased from the butcher and sent back to Windsor, a 'royal pensioner' for life."

Naturally, a royal ox is a famous ox, and the nameless beast (generally referred to as the Windsor Ox) had his portrait painted, and prints of the painting circulated throughout the empire. Hugh Watson's daughter recounted that a young acquaintance of her father, while on a shooting expedition to the most remote corner of northeastern India (Pakistan's territory today), happened upon a small adobe house (which she called a mud hovel). The hunter entered and found "its only occupant . . . engaged in his devotions, prostrate before a rude altar, over which was suspended a coloured print" of a very contented ox. When the young man examined the print closely, "he observed it was labeled 'A Polled Angus Ox, the property of H.R.H. Prince Albert, and bred by Hugh Watson, of Keillor.'"

The Keillor ox was the sum total of royal interest in black polled cattle until some thirty years later when, after Albert's death in 1861, the queen began spending the temperate months at Balmoral Castle in Scotland. By that time William M'Combie was the owner of the premier breeding herd, housed at his farm at Tillyfour. Hugh Watson's herd had been dispersed in 1861, the year after his death.

M'Combie was a successful showman, both in fat-stock and breeding-stock competitions. Aside from his triumphs in Paris, his most notable success and the greatest publicity came from his fat steer Black Prince. This enormous beast, which had an estimated live weight of some 2,500 pounds (1,135 kilograms), an unlikely but possible heft, swept the classes at the Birmingham show in 1866 and then took first in the bullock class at the Smithfield Christmas show after a brief detour for a command performance.

Having heard of this notable animal, Queen Victoria requested that it be brought to her at Windsor Castle. M'Combie took it to Windsor on its trip to Smithfield; the queen admired it enormously. She may have directly, or by demeanor, indicated an interest in how such a huge animal would do on the table. After the Smithfield victory, M'Combie sold it to the London butchers Lidstone and Scarlett, purveyors to Her Majesty. He retained, however, the "baron" roast from the retail market and gave it to Her Majesty as a Christmas gift from Tillyfour.

This was not an insubstantial gift, but it was appropriate for a queen. The "baron" roast is where the very best cuts of the animal are taken. Kept in one piece, the baron is a cross section of the most tender and flavorful meat taken from the rump and the ribs. It is alleged that such a chunk was cut and spit-roasted for Henry the Eighth and that he so admired it he elevated it on the spot to the rank of Baron Roast. In steak-house terms, the baron roast would contain all of the porterhouse, T-bone, strip, sirloin, and tenderloin. Given Black Prince's live weight of 2,500 pounds, the steer should yield a baron of some 144 pounds (65 kilograms).

It was not to be the queen's last enounter with Black Prince. Two summers later, while she was in residence at Balmoral, she visited M'Combie (who was also by then a member of Parliament) because she was considering establishing a herd of black cows at her Scotland home. On entering the dining room she saw and recognized the head of Black Prince mounted on the wall above the dining table.

After lunch, according to local legend, M'Combie gave a command-performance cattle show for her. While she sat inside a new wing on his house, one built specifically for her visit, M'Combie's herdsmen paraded cattle past the window. His herd had been decimated by pleuro-pneumonia in the year before Victoria's visit, and so, again according to local lore, he paraded the same animals past her several times, the herdsmen changing shirts and caps while out of the queen's sight.

"According to local lore" is no casual figure of speech. While in Scotland, the author was taken by a well-known Aberdeen-Angus breeder, Willie McLaren of Blackford, on a drive to some of the great historic farms. We came to Tillyfour on a spring afternoon when the hillsides were truly carpeted with the bluebells of Scotland. There didn't seem to be anyone home, so we stood in the drive and admired the simple Georgian stone house. The resident, who turned out to be an oil-exploration company executive (the nearby port cities are gateways to the North Sea oil fields), saw us and came out to chat. Now here is this Englishman, new to the area, and he had already been imbued with the story of Victoria and the charade parade and told us about it with amusement. He added an item about the event that is not part of the usual account: The east wing of the house was the one constructed at the last possible minute specifically for Queen Victoria's inspection of the herd, he explained, and added the detail that had escaped our notice. The rear of the new wing, which faces south and would ordinarily be used to pick up solar energy in somewhat dark Scotland, is entirely deficient in windows. There was no possibility that Victoria, viewing the animals through the north-facing windows, could have seen the herdsmen repeatedly changing clothes before leading animals around the house. It was a gentle deception: One would not like to disappoint a guest, especially a royal guest, who had traveled some distance to see a herd of cattle.

From the point of view of all the breeders of Aberdeen-Angus cattle, her visit was a rousing success. The following year she established a herd of registered animals at a farm in royal ownership near her summertime retreat at Balmoral Castle, drawing from M'Combie's stock and others. The cachet of a royal association is no small thing in Britain, where "by appointment to Her Majesty the Queen" can still improve the sales of everything from marmalade (by Wilkin & Sons, Ltd.) to mole control (by Mole Control & Pest Services).

. . .

If, by some happenstance of taste, Victoria had chosen to build a summer castle in western Scotland, there is every chance that she would have picked a different breed, for Scotland had produced two notable beef animals. The other, and it is a goodly cow, is the Galloway. It too is all black (originally), and it is beautiful. It too has given up horns, and it has a disposition noted for amenability. This is always relative with cattle, but the Galloway is fairly easy to handle most of the time. The cows do have a tendency to be overprotective of their calves, even more so than Aberdeen-Angus. In spite of its many excellent qualities, the Galloway's failure to become a worldwide favorite like the Aberdeen-Angus highlights, by contrast, some of the most valuable commercial traits in the sleek black cows.

Galloway is a rural district on the west coast of Scotland that comprises the southern portion of a large bulge sticking out into the Irish Sea south of Glasgow. On its native turf, the Galloway cow is all black, but most of the breed seen abroad are black-and-white cattle. They look approximately like an Aberdeen-Angus with a white sheet tied around their middle. These are the Belted Galloways, extremely striking animals raised with considerable pride by owners who want a truly eye-worthy animal. The white, always confined to a single wide band around the cow, is a nineteenth-century addition to the all-black hide. Whether on pigs, pigeons, chickens, or cows, bands were highly prized by aristocratic landowners. This fad is of an entirely mysterious origin. Although many authors comment on the deliberate breeding of banded or belted animals, no one seems to have the faintest clue as to the reason.

Belted Galloways should not be confused with the most widely distributed black-and-white cow in the world. That would be the horned dairy breed of Holstein-Friesian blood. The Holsteins wear random splotches and blotches of white—the prototype for a million suburban mailbox paint jobs.

A Belted Galloway cow

There is no typical Holstein, either in coat pattern or size. Most of them are among the largest of domestic cattle, but a smaller cow with a good milk output will be found in dairy herds.

But there was a time, less than two hundred years ago, when every cattle breeder, feeder, and butcher knew of Galloway, for its beef animals were the standard of perfection. Youatt, in his *Youatt On Cattle,* thought the Galloway utterly superior to the Angus. A proper Galloway, he wrote, "is longer than . . . the Aberdeen-Angus" where it mattered, "long in the quarters and ribs, and deep in the chest. . . . There is less space lost between the hook or hip bones and the ribs than in most other breeds." This was "a consideration of much importance, for the advantage of length of carcass consists in the animal being well ribbed home, or as little space possible lost in the flank." The legs, Youatt thought, were unimprovable: "The Galloway is short in the leg, and moderately fine in shank bones—the happy medium [is] preserved in the leg . . . there is no breed so large and muscular above the knee [hock]."

It should be noted that Youatt is writing before the first two great breeders of Aberdeen-Angus, Thomas Watson and

William M'Combie, owned a cow. Perhaps more important, he is writing in that time when Scotch cattle were still being driven down to England to fatten before going to the butcher. He noted that the Angus did very well on its own turf but suffered by comparison when it and the Galloway were both sent south to finish. There, the Galloways thrived.

Youatt was not entirely wrong to discount the eventually successful Aberdeen and Angus cattle. They did not come into universal favor until the better breeders, like Watson, M'Combie, and several less-famous men, finished their own cattle and brought them to the market and to the agricultural shows in best condition. What the Aberdeen-Angus men were trying to do (as were the English Shorthorn breeders), the Galloway man had already achieved.

The Galloways, Youatt summed up, "are native to the country and incapable of improvement. The intelligent Galloway breeder [one who would agree with Youatt] is now perfectly satisfied that his stock can only be improved by adherence to the pure breed, and by care in the selection."

William M'Combie, the only founding father of the Aberdeen-Angus breed who left us an autobiography, was hardly disparaging of the curly-coats from across the country. While the Aberdeen-Angus had a reputation for doing well on nothing but grass, M'Combie knew that the Galloways were excellent for grazing on second-rate pasture. They would fatten, he wrote, on ground where his Aberdeen cows would waste away. They had that great quality of "thriftiness."

As noted earlier, the Angus were thought to be one breed, the Aberdeen another, and the Galloway the third of the black hummlies. The first breed book, that is, the record of parentage and ownership of registered animals, included all three, only putting an asterisk by the curly-haired Galloways to distinguish them from the sleekies. After decades the Polled Cattle Society eventually became the Aberdeen-Angus Society of Great Britain. Even after the Galloway breeders went their own way, the breeds were shown together in the same class at the great competitions, a situation that did not change until 1891.

The first history of the Aberdeen-Angus breed, by James Macdonald and James Sinclair, was published in 1882, while the question of whether Aberdeen-Angus were one breed or two was not yet settled. Their volume is somewhat cautiously titled *The History of Polled Aberdeen or Angus Cattle*. But whether one type or two, Macdonald and Sinclair went to excessive lengths to deny that there is any Galloway admixture to the Aberdeen-Angus cattle.

For almost twenty pages the authors disparage the possibility that there could be Galloway blood in the registered black sleekies. But in those days, as the breed was being improved, any number of bulls and cows were registered whose parents were not themselves registered. This makes perfect sense, of course. You have to start somewhere, and until registered stock is scattered through the universe of black cows, the odd local black polled beast will sire or calve an Aberdeen *or* Angus that ends up in the breed book. So a denial of Galloway influence was as much an act of faith as reason. The only factual argument offered that there is no Galloway influence in the Aberdeen-Angus gene pool is the experience of Lord Panmure, a famous black sleekie breeder who tried Galloway bulls on his herd and found them lacking. This argument is repeated by the authors twice within thirty pages. The authors protest too much. And they also offer some contrary evidence.

Some breeders evidently admixed the curly-coated Galloways. Up in Aberdeenshire, near Huntly, the Duke of Gordon had no qualms about interbreeding any and all of the black polled animals. His herdsman is quoted in the official history to this effect: "The present Duke of Gordon has at different times within the last 30 years [1800–1832] brought the best selection of bulls and cows that could be found in Galloway into the district . . . from which great benefit has arisen, by their increase and mixture with the original stock and with each other; and his Grace's example has been extensively followed by agriculturalists and breeders of cattle." This is so contrary to the earlier assertions of MacDonald and Sinclair that one suspects them of deliberate waffling. Working for the Aberdeen and Angus breeders who made up the bulk of the society members, they may have been obliged to argue that the sleek ones were entirely unrelated to the curly-coated Galloways, but for their self-esteem they also quoted this earlier report to the contrary.

The Galloway, with its calm manners, was also popular with recreational breeders and, when that mysterious fad of belted farm animals sprang up in the mid–nineteenth century and the striking Belted Galloways were created by selective breeding, one suspects the addition of some cow with the crude beginnings of what would become the band. In other words, the mixture of genes may have gone in the reverse direction, for a likely choice is an old-fashioned Aberdeen-Angus of the type that regularly was white along the belly, a good starting point for the breeder seeking a circumferential band of white. There were and there still are rare occasions when a belted Aberdeen-Angus appears in a British herd where white has not been entirely eliminated. If so, it has always been neutered, fattened, and eaten. There could also have been a breed of belted Herefords—that white-faced, red- and white-haired cow occasionally brings forth a belted beast of considerable visual appeal—but it too would not have met the breed standard and would have been consumed.

It should come as no surprise if some future genetic testing indicates that Galloways are intermixed with Aberdeen-Angus.

This would happen not only by design but by the long-standing practice of cattle stealing, something of a Scottish national industry until the Tudors brought some semblance of law and order to the Highlands in the eighteenth century. Cattle would be stolen on the West Coast and driven east toward Angus. With the fortunes of clans waxing and waning, Angus and Aberdeen polled cows were certainly liberated and moved across the lowlands to Galloway.

A glance through any collection of photographs of Aberdeen-Angus cattle will turn up some suspiciously hairy beasts, and these are not just photographs taken in the dead of winter when even the sleekest blacks put on some winter hair. There are animals in breeding-stock herds with persistent curly coats over parts of their anatomy (often over the shoulders in bulls and the rump in cows). And there are some smallish Aberdeen-Angus, particularly in Australia, that look more like Karakul sheep than black hummlies.

Hairiness was not considered a fault in the early days. As we noted earlier, William M'Combie purchased a cow from Mr. Fullerton that would become the founding dam of the famous Queen and Princess lines. The dam of that cow, retained by Mr. Fullerton, was registered as an Aberdeen (before the breed was combined), Black Meg. Here is Fullerton's description of that cow, written years later: "Then her hair—my eye, such hair!—we shall never see the like again; [hair] of the best quality, and on to her flanks you could almost hide your hand in it." That sounds suspiciously like an animal with some Galloway blood.

But the Galloways had faults, and there are reasons the world has pretty much passed them by in favor of Aberdeen-Angus. First, the hair: When black polled cattle are kept penned up for the winter, as they commonly are in eastern and northern Scotland, the ordinary amount of dung rubbed into a winter-coat Aberdeen-Angus is a problem. The Galloways would be a disaster if held and fattened in stalls. In their milder coastal native shire, grazing on extensive unimproved pastures, they lived outdoors year-round.

The same problem would happen in extensive agriculture as it is done in North America or Australia. There, lean steers and culled heifers go to feedlots for up to three months of fattening under crowded and messy conditions. The Galloways with their perpetual coat of longer curlier hair would be at a distinct disadvantage: susceptible to disease and difficult to clean before being slaughtered and skinned out.

Another drawback is that first-generation crosses between Galloways and other breeds do not seem to produce the same amount of hybrid vigor as can be expected when Aberdeen-Angus are crossed with other beef breeds. This is not a fault if, as Youatt said, the pure animal is hardly capable of improvement. It would be a problem, however, for someone interested in making money selling purebred beef sires. By far the largest number of Aberdeen-Angus bulls, rather than being kept as sires in purebred herds, are sold to improve the beef quality in other breeds by crossbreeding.

Farmers who want a small herd of amenable and delicious beef critters, ones that are finished on the farm and custom-slaughtered, are justifiably happy with and proud of their Galloways. They may not be commercial cows, but let William M'Combie have the last word: The Galloway, he wrote, will succeed on pasture "so poor that our Aberdeens could not subsist on it. . . . You can bring them to be three-quarters fat, and there they stick; it is difficult to give them the last dip. If, however, you succeed in doing so, *there is no other breed worth more by the pound weight than a first-class Galloway.*" (Italics in original.)

→ 6 ←

Angus and Aberdeen
Come to America

MONEY, nothing more romantic or idealistic than that, brought the first Aberdeen-Angus to North America. The rise of manufacturing centers, the concentration of wealth in the cities of the victorious northern states after the American Civil War, and, most of all, the sudden appearance of an urban middle class with disposable income created an insatiable market for all kinds of luxury goods, including luxurious food. Before the Civil War, there was practically no such thing as a restaurant. Food was served in hotels and inns; it was assumed that the only people not eating at home were travelers. For the single men and transient workers, a boardinghouse would have to do. Until the second half of the nineteenth century, cookbooks were merely domestic bibles of thriftiness. But the famous Delmonico's Restaurant in New York City (which had started out as nothing more than a candy store) was open and thriving by the middle of the Civil War, and after the war the first cooking schools for housewives (as opposed to schools for domestics) were founded. Fannie Farmer published her first cookbook in the 1890s, based on her Boston Cooking School curriculum. Among other things, it is a document in high praise of quality beef, always to be cooked rare. The era Mark Twain christened The Gilded Age had begun, and the beef industry reached for a share of the wealth.

Black cows from Scotland arrived in the United States relatively late in the nineteenth century, at the height of the post–Civil War boom. English purebred Shorthorns made it across the Atlantic before the Civil War and became the primary breed for upgrading American stock by the 1860s. In relative timing, this follows the same sequence that had occurred a few decades earlier in Britain. There, the commercial boom in registered breeding stock began with the success of the Collins brothers' Bakewell Shorthorns during the first few decades of the nineteenth century, followed by the triumphs of the Aberdeen-Angus both in Britain and on the Continent toward the middle of the century.

. . .

The first Aberdeen-Angus to come to the New World arrived in Canada in 1861, a dozen years before the first black cows arrived in the United States. They were two registered Angus (using the terminology of that era when the registrations of Aberdeen and Angus were kept separate) consigned to Sir George Simpson, governor of the Hudson's Bay Company. Simpson was the most important, or prominent, private citizen in Canada, a colony then that was barely settled west of Ontario and north of Quebec City. What artifacts of civilization there were in remotest Canada arrived in the stores and trading posts of the Hudson's Bay Company, by far the largest and most far-flung economic force in the colony. For reasons that are now wholly obscure, Simpson had made the acquaintance of the

Earl of Southesk, who shipped him a heifer named Dorothea and a bull called Orlanda. The Lord Southesk obviously intended that Simpson establish a herd of purebred black doddies in Canada. It is not possible to overestimate the intended value of the gift. The Southesk herd was one of the greatest sources of breeding stock in Britain, until it was destroyed by rinderpest. It would be so even if it produced only one animal, the great cow Erica.

Unfortunately for everyone involved, including the animals, Governor Simpson died shortly after the cattle arrived, so there is no record of what happened to them. If they procreated, no offspring were recorded in the then Polled Cattle Society. (For many years, even after there were Aberdeen-Angus societies in North America, stockmen from the United States and Canada would register their animals with the British society.) One suspects the two Southesk Angus were eaten. On first sight, persons unfamiliar with the polled black sleekies always thought there was something wrong with them. They had no real color, just uninterrupted blackness, and no horns.

Lord Southesk's gift notwithstanding, the introduction of black polled cattle to North America is usually traced to 1873 when several black polled bulls were sent to a short-lived experiment called Victoria Colony in central Kansas. This primacy is endorsed by both the American Angus Association and the Aberdeen-Angus Society of Great Britain. The first printed reports of their arrival announced that they were Galloways, but that mistake is understandable because in those years, as noted earlier, the Polled Cattle Society's herd book included Galloways.

"Among the more noticeable importations," wrote a western stockman, "are a number of black hornless bulls of the Galloway blood, which have all the beef qualities of the Durham

[Shorthorn], maturing fully as early, and possessive in addition of habits of industry and are extremely hardy and thrifty."

Grant, the patron of the Kansas colony named for his queen, was a wealthy man who had some trouble keeping his money but no trouble making more. His most memorable financial coup has a peculiar and morbid connection to the Aberdeen-Angus breed. Prince Albert, Queen Victoria's adored consort, had kept a few Aberdeen-Angus around the castle at Windsor (including the famously rescued ox that licked his hand and got a royal pardon). In 1861, when Albert fell ill with cholera (bad drains at Buckingham Palace are thought to be the cause), George Grant had an inspiration. On hearing of the consort's illness, Grant bought up every yard of black crepe in London. When Albert died, Grant turned a fortune by exploiting his monopoly, overcharging the bereaved citizens for the obligatory outward and visible sign of inward and spiritual grief. Not a draped window, a bedecked carriage, or a black-scarfed Londoner mourned Albert without adding to Grant's bank account.

The first history of the Aberdeen-Angus Society of Great Britain mentions that Grant's three imported animals became immediately notable as sires of vigorous hybrids when they bred local mixed-breed cows. Purebred Shorthorns were already being used for upgrading range cattle, so there was a basis for comparison. One American writer is quoted as claiming that the "polled crosses did best, standing the winter far better than the Shorthorn crosses. The polled crosses weighed about 120 lb. live weight (54 kg) more [at the time of sale] than the other crosses." That is probably the strongest evidence that the original animals were from Aberdeen or Angus, since the otherwise excellent Galloways do not produce a great deal of hybrid vigor when crossed with other breeds.

The Victoria Colony disappeared within a few years, and

Grant's Aberdeen-Angus bulls, having no purebred cows with which to continue the breed, simply scattered their seed into the broader genetic pool of North American cattle, leaving no traceable heirs. A single bull was sold to a farmer who liked it well enough that a few years later he imported male and female registered breeding stock from Scotland. His Victoria Colony bull seems not to have played a recorded part in that herd's genetics; there is no mention of it in the breed book. The utter disappearance of the Victoria Colony bulls is not unlike the fate of the Roman soldiers guarding Hadrian's Wall on the Scotland-England border. Rome fell to the vandals, the mails failed, the payrolls were not forwarded, and the legionnaires, most of them from the Near East, merged into the population of Cumbria.

Although Grant's colony disappeared just as fast as the Aberdeen-Angus cattle he imported, there is still a town of Victoria, Kansas, and on the centennial in 1973 a delegation of British Aberdeen-Angus breeders journeyed there to read a proclamation from Queen Elizabeth II in honor of, in no particular order of importance, the cattle, the American Angus Association, and George Grant. There is a statue there in Victoria, a modest marble column with a small bronze Angus bull on top.

William (known to all as Willie) McLaren of Blackford, Scotland, proprietor of the notable Netherton herd, read the queen's proclamation. He also had a speech prepared of greetings from the British Aberdeen-Angus Society, which he was not given time to deliver. Why the American Angus Association did not want to hear a speech from the man who is one of the finest breeders in Great Britain is inexplicable. Anyone who has spent some time in Scotland with Aberdeen-Angus Society members and knows their hospitality understands that if an American stockman was invited across the Atlantic with a presidential proclamation and a brief speech in his pocket for the Aberdeen-Angus Cattle Society of Great Britain to hear, not only would the American guest be allowed to read both but he would be applauded vigorously and afterward have considerable trouble paying for drinks at the bar. There are times when it seems that more than an ocean separates the English-speaking nations. An American Angus breeders delegation to the World Aberdeen-Angus Congress in Scotland is remembered with some bemusement for insisting on having its own Americans-only buses and its own itinerary of herds to visit.

The Aberdeen-Angus Society of Great Britain still maintains contact with the royal family, even though the Balmoral herd was dispersed in 1934. The late Queen Mother was the patron of the society until her death in 2001, and her grandson, the Prince of Wales, someday to be King Charles III, has a commercial herd of Aberdeen-Angus. Like all his Cornwall estate products, they are being raised most organically, a role which they are better suited to than the average cow.

Sixteen years later, at just about the same time that George Grant's Victoria experiment failed and his nonregistered Aberdeen-Angus bulls disappeared from sight, the very first successful breeding family of Aberdeen-Angus cows arrived in North America. In 1876 William Brown, a Scot and superintendent of the Guelph, Ontario, Experimental Farm, imported a yearling bull, Gladiolus, number 1161 in the British breed book. (Gladiolus sounds a bit twee for a heroic bull, but the British have an odd way with flower names. During World War II their fleet of brave and adventurous antisubmariners served in Corvettes with floral names like HMS *Asphodel.*) On January 12, 1877, a cow named Eyebright (3001 in the breed book), who was shipped already bred on the boat with Gladiolus, gave birth to a calf, Eyebright 2nd (registered as number 7444 in the breed book), and she is unquestionably the first purebred Aberdeen-Angus born in North America. By the standards of the time Gladiolus and Eyebright would have been registered as Aberdeens; they were both bred on the Earl of Fife's estate near Banff. In the next few years Eyebright 1st and Gladiolus produced more offspring, and additional animals were imported to the Guelph herd. Besides being the first born in North America, Eyebright 2nd was also the first purebred cow exported from Canada to the United States. She ended her days at Kansas State College (now Kansas State University) in Manhattan. The Guelph herd is no longer in existence; like so many college and state extension-service herds, whatever breeding occurs is done with artificial insemination and all the research is into crosses for the commercial meat industry.

Persons studying the pedigree of black cows will recognize that the Guelph herd brought all manner of famous Scottish bloodlines into Canada, from where they entered the United States. Persons not studying cow ancestry can at least admit that Canada, for all its modesty, is the true mother of Aberdeen-Angus in the New World.

In 1878 the first herd of purebred Aberdeen-Angus (one bull and five cows) found their way to the state of Illinois. The proximate cause, nay the primal cause, of this event was money. Two Chicago-area Scots imported the second breeding herd into North America after learning that William M'Combie was winning practically every show he entered (including sweeping the Paris Exposition for the second time in 1878) with black cattle, garnering cash prizes and gold medals when gold medals were 24-carat.

James Anderson and George Findlay of Lake Forest, Illinois, formed a partnership to import Aberdeen-Angus when the state of Illinois announced plans to hold a grand exposition at the Chicago stockyards, with huge prizes for beef cattle. Hoping that black sleekies could be as dominant in the United States as they had become in Great Britain and on the Continent, they seized the opportunity. Findlay placed an order with his brother, then living in Scotland, to ship him some winners. Shortly, there arrived the bull and his five cows.

The Anderson/Findlay herd took prizes at local fairs and at the first Chicago Exposition in 1879, as well as the St. Louis Exposition and the Illinois State Fair. Grand championships and huge cash prizes eluded them, but their results were nonetheless outstanding. That they won anything was a tribute to the animals, as the American judges had never seen purebred Aberdeen-Angus cattle. There are (and were) some subtle differences between what constituted prizewinning form for Aberdeen-Angus and for their principal competitors, the Shorthorn and the Hereford. A proper Aberdeen-Angus butt, for example, is rounded off on the edges, whether you are looking at it side-on or tail-on. Proper Shorthorns and Herefords should have squarer, boxier cabooses. It was amazing the black cows did so well, considering that Aberdeen-Angus entries were something akin to entering a tulip in a daffodil contest.

The Illinois Aberdeen-Angus herd's success sparked an interest in black cattle that increased steadily through the 1880s, slowed but not stopped by the crash in cattle prices in the nationwide recession of '88. Not only did the hard times hit cattlemen financially (western ranchers carried more mortgage debt per capita than any other group of citizens in the United States), but the recurring blizzards of '86–87 and '87–88 put any number of potential buyers out of business. When the plains recovered toward the end of the century, Aberdeen-Angus cattle once more became desirable stock for upgrading western herds. From such small beginnings, from mere handfuls of stock, the Aberdeen-Angus today produce more beef in North America than any other breed. Not all is purebred, of course. The same thing is true today as was seen in George Grant's first animals. When bred to another breed (or to an Angus cross anything hybrid), no animal exceeds them in producing beef. As they say in Scotland, where they are as plainspoken as they are hospitable, the black steers "die well."

Angus on the High Plains and at the Felton Ranch

THE Montgomery family and thousands more came into Montana shortly after the blizzards and depressed prices of the 1880s when a second east-west railroad, James J. Hill's Great Northern Railway, opened up the country north of the Missouri River to homesteading and town building. Even though a mighty river runs through it, the countryside along the High Line (as the Great Northern was nicknamed) couldn't get much drier. The Missouri cuts deep into the land and waters only the floodplains at the bottoms of its canyons. People homesteaded on places with names like Sand Springs and Big Dry Creek. By the time the homesteaders got there, what water there was had largely been acquired, one way or another, by large ranches. What brought the last homesteaders into Montana (and large portions of North and South Dakota and Wyoming) was a popular belief, encouraged by the railroads, that "rain followed the plow." Of course, it turned out that it didn't, and as one homestead after another was abandoned, eastern Montana got the nickname it deserved—The

Big Dry. But land that couldn't be farmed because it was too high to irrigate from the region's few rivers could be grazed. It might take 200 acres (80 hectares) to feed a cow and her calf, but if the rancher could find a little water from a spring or make enough water flow with the ubiquitous windmill, raising cattle was difficult but not impossible.

My cousins who own the Felton Angus Ranch didn't start out to be Aberdeen-Angus breeders. Raymond Felton and his wife, Margaret (née Montgomery), began with a farm on Nine Mile Creek, a small tributary of the Clark Fork of the Columbia, several miles downstream (and north) of the university town of Missoula. They started out there shortly after World War II, running a small dairy (with handsome Jersey cows) and a few Shorthorn beef cattle on leased land in the neighboring National Forest. As a business, the ranch is still in the same state but has moved east, with two ranches—one along the Yellowstone River; the other, larger spread on the Tongue River south of Miles City. The world of the Clark Fork on the western side of the Great Divide is just about the exact opposite of the High Plains. The Clark Fork valley is technically a maritime climate, although one that can see double-digit below-zero winter days. The main difference is that it can rain too much in a maritime climate. It never rains too much on the east side of the Rockies. The Feltons have moved twice over those five decades, and by coincidence their movement from the western slope of the Continental Divide out into the plains replicates, almost to the letter, one half of the history of beef cattle in Montana.

Twenty years before the first Texas cowboy pushed big herds of longhorns onto the High Plains, small farms were scattered up and down the mountain valleys of western Montana. There were cattle on Nine Mile by 1870, providing butter and meat to the little gold-panning boomtown of Stark at the head-waters of the creek. By Montana standards, that is ancient history, and the settlement of Nine Mile is as remote, in the mind, as getting off the *Mayflower* at Plymouth Rock.

The combination of animals on the Felton place on Nine Mile was typical of Montana ranches from the mid–nineteenth century through World War II. However, except for a few religious colonies or weekend ranchers, ranches no longer raise a variety of livestock. Compared to today, the first oddity was the Feltons running dairy and beef cattle on the same farm. There were chickens and pigs as well. The hogs were given gallons of milk after most of the cream had been centrifuged out. (For the first several years on Nine Mile, the ranch sold only cream, which even unrefrigerated and soured would make perfectly good butter when it reached Missoula.) A few sheep bumbled around near the cow barn. They seemed to be more or less pets, since they were well past the delectable lamb stage.

Anyone wanting to experience the Old West won't find it on a dude ranch or even on a working stock-raising operation. To see what pioneer ranching really looked like, find one of the antimodern religious settlements, Amish or Hutterite, scattered across Montana. Just up the Tongue River valley from the Felton ranch there's a new Amish settlement—they came into the country in the mid-1990s—that really does resemble Montana in the 1800s. They cut hay and plant seed behind teams of horses. Their lamps are kerosene, not electric. They go to town in buggies. The Feltons had just moved onto their Tongue River ranch when the Amish showed up at theirs. Cousin Maurice, that's Margaret and Raymond's oldest son, being neighborly, gave them a few tons of horse-drawn farming implements that were no longer used around the ranch. Even after a farmer buys his first tractor, he will hang on to the horse-drawn stuff. Most of it could be pulled by a small tractor, and it survived for decades after the last draft animals died. Throwing away something that might work or might have a useful piece of tool on it is not done. On average, farmers and ranchers and dairymen have much more scrap metal per capita than any other occupation (except junk dealing). They are weighted down, but the legendary cowboy drifted along, as the song goes, as light as a tumbleweed.

In the early days, beef and lamb on the hoof and pork on the trotter were all imperishable and transportable as were butter and eggs, which both have a reasonable shelf life without refrigeration. The small towns and mining camps provided markets decades before the railroads arrived in the West, before the rail lines provided the shipping that in turn lured those Texas herds to the High Plains. Besides the self-sufficient Amish, there is just one ranch along the Tongue River valley that still has pigs, chickens, and milk cows to go along with the beef critters. It belongs to a young couple who live way back off the road, doing subsistence farming in addition to running some beef. Ranching has gotten pared down to a few repetitive occupations.

There was a plaintive help-wanted ad in the *Miles City Star* a few summers ago that summed up the distinction between cowboys way back when and ranch hands now. A big outfit in Wyoming wanted two men who were willing to irrigate fields and put up hay, "in addition to working with stock." The advertisement promised excellent pay and two unheard-of benefits for ranch hands—medical insurance and paid vacations. "This will be a good job for a cowboy who will farm, or a farmer who likes to ride," the ad explained. That same advertisement, absent offers of insurance and vacations, could have run in a frontier newspaper one hundred twenty years ago when the first fence-building, hay-making cattlemen came into the country.

The Feltons got into the registered Angus business slowly, with one major detour or diversion at the beginning. Margaret and her husband, Raymond, had decided on two major improvements. They would get some black into their Shorthorn beef herd with their first Aberdeen-Angus bull and step up from selling soured cream to selling Grade A milk. Both projects were upgrades, as they say these days. The author helped pour the concrete floor for the milking room during a summer visit. That is ranch life. If you can help, you don't wait for an invitation; if you can't, you stay out of the way. Raymond Felton, like most ranchers, thought that a boy, even a visitor, could run anything if he was tall enough to reach the controls. That includes horses. If he can get on it, he can ride it. If he can see over (or through) the steering wheel, a boy on a farm can drive the pickup. If he wants to go shoot something, he can borrow a gun. This is all equally true for girls. For a skinny suburban kid just entering adolescence to be treated like a grown-up is a remarkable tonic and never forgotten.

The trouble with dairying is pretty simple, and it reinforced Raymond Felton's interest in beef cattle. There are no days off for a dairyman, and all the other chores must be fitted into the gap between morning milking and evening milking, everything from haying to irrigating to feeding to fencing. Cleaning up the Grade A parlor and the animals takes hours more than just getting them into the old barn, wiping down their udders, and hooking them up to the milking machine. Beef cattle, even if they spend most of the summer taking care of themselves, need a great deal of attention at breeding season and calving season. There are shots to give and ear tags to put on and records to keep. All that just doesn't fit neatly into the hours between milkings. It's easier in some ways to have a full-time eight-hour job and raise beef cattle than to try ranching while running a small dairy. The part-time rancher has a schedule for the day job and gets weekends off.

Fortunately, Raymond got out of dairying and seriously into Angus cattle before those sweet, mild-mannered, good-looking Jerseys wore him into the ground. If you see even a single milk cow on a ranch these days, it will be some kid's 4-H project. Cattlemen will do ditches, mow fields, grind feed, build fences, and run sprinklers, but they won't milk cows.

The amount of land available for ranching has been dwindling at an alarming rate in Montana during the last two decades. Any land within commuting distance of a market center or a city becomes worth too much for cows to walk on it. The Feltons sold their Nine Mile land (just twenty-five miles from the center of Missoula, home of the state university) to a developer. It wasn't much of a development by today's standards, but it raised enough money to move the Angus operation over the divide to the Yellowstone Valley between Livingston and Billings. Instead of the Forest Service, they leased their summer high pastures from retired ranchers and intensively farmed their own property down in the valley. It seemed ideal. There was land to rent and plenty of irrigation water in the Yellowstone, plus good wells in the foothills, and no arguing with the Forest Service about how many cows and calves could be put out on leased public pasture. They were so far out of any town that they got their own exit on Interstate 90. (You see ranch exit signs only in fairly narrow valleys where the limited-access interstates have to run where the old highway was. Other than those situations, the old highway is left in place as a frontage road between exits and serves the local landowners as it always has.) The boys, Maurice and Richard, grew up and got married, and then there were Felton grandchildren. The registered Felton Angus Ranch stock sold well and paid all the bills.

The devil has many disguises. In the case of the Felton ranch on the Yellowstone, it turned out to be a movie about fish. This is absolutely not an oversimplification. The movie was *A River Runs Through It,* and it convinced a number of people who had too much loose change that the Most Important Thing in the World was to own a ranch with a river that ran through it with trout swimming around—and it had to be in Montana.

What the people with excellent cash flow craved was quiet, privacy, and famous neighbors. That ruled out the Felton ranch. Although like so many on the Yellowstone downstream from Livingston, Montana, it had superb fishing, it also had the four-lane divided highway and a very busy railroad line along the river. No one would go to a movie called *An Interstate Runs Through It* or *Brown Trout by the Burlington Northern.*

The outsiders descended on the Boulder River drainage just east and over the hill from the Felton Angus Ranch. The Boulder River is a major tributary of the Yellowstone running through a quiet and out-of-the-way valley where real-estate agents can promise privacy. It has pretty good fishing and runs next to two-lane roads. To the newcomer, privacy always means No Trespassing signs and usually includes kicking the cows off the place. Unfortunately, the divide between the Yellowstone and the West Boulder was where the Feltons had rented their summer pastures. When their leases could not be renewed with the new owners at the end of the 1990s, it was time to do what any sensible rancher would do—sell some of their Yellowstone property on the north side of the river, as far as possible from the interstate and Burlington Northern tracks, to a developer and move most of the cattle to a place that was so far from trout water that no one would bother them. And they bought all the summer pasture they needed, so they would never worry about *that* again. They kept the Yellowstone hayfields and enough of the foothills south of the river to run a small part of their operation at Big Timber. Because it is more central than the Tongue River ranch, they still have their annual production sale there. The sale barn at Big Timber is just fifty yards off the interstate, while an auction on the Tongue would be fifty-five miles, forty of it gravel.

The pressure continues on Montana cattle ranges even in economic downturns like the turn-of-the-century dot-com crash. Ranches look safer than the stock market and sexier than suburban real estate. Working ranches with trout water are either being subdivided into cabin lots and ranchettes or sold in toto if someone has a roll of at least three million bucks that's burning a hole in his pocket. In the first half of 2003, all of the ranches sold in western Montana (meaning in trout country) brought in prices starting at three million dollars for a thousand acres, and on up from there, and at least 90 percent of the sales were to out-of-staters looking for a second home. Subdividers offer ranchettes (usually twenty-five acres) at fifteen thousand dollars an acre and up. Prices of one thousand dollars an acre are being asked for unimproved (and unimprovable) High Plains land, so barren that it takes a hundred acres to feed a cow and her calf through the summer. If someone wants to hunt antelope or pheasants bad enough, prices that high and higher are being paid out in the Big Dry. The math is simple. It would take as much as a hundred thousand dollars (or a bank loan carrying a twelve-thousand-dollar annual payment) to buy enough rugged and dry High Plains to feed a couple of cows and a calf for just five or six months out of the year. And after feeding that calf hay through the winter, a rancher might get a thousand dollars for it at the Miles City auction yard. These numbers do not work for a young man looking to get into the cow business. Even without trout (and most of Montana is in a trout-free zone), there's always hunting, and more and more lightly watered High Plains ranches are sold each year for various shootist purposes.

All this spare-cash recreational real estate makes it impossible today to replicate the Felton experience in Montana, and probably anywhere from Arizona to the Klondike. No one can buy in small and grow. When the Tongue River ranch gets sold, and it will someday, there are a couple of choices. Most large working cattle ranches in the West are now owned by multimillionaires or the multimillion-dollar corporations they control. One of these bigger ranches will buy it and add it to the tax deductions. And the other choice is someone interested in hunting or, as does happen occasionally, a person or a sect with an extreme desire for privacy.

.   .   .

Montanans may say of the Big Dry that it is hard country. That is not to be taken by a visitor to mean something negative; it's just the fact. The great drought of the 1920s combined with a fall in commodity prices brought the Great Depression to the West long before it hit the thickly settled and urbanized parts of the country. With that ten-year head start, what people from the plains call the Dirty Thirties was just the final blow that drove the dirt farmers and the small ranchers out of the country. The ones that stayed knew exactly what they were getting into, and they chose to do it.

Years ago I went looking for my grandfather's long-abandoned ranch north of Chinook, Montana, and made the mistake of misinterpreting what "hard country" means to those who survived and stayed. The man from the land title company in town (surely the only person with a clue to the location of a ranch sold off in the 1930s) was kind enough to drive me out to see where the ranch had been. As we turned off the straight-line section road, he remarked that we were almost there, but before we arrived, he said, the land would get harder. And when it did, when more rocks and prickly pear cactus appeared on the surface and the clumps of bunchgrass got farther and farther apart, I said, "You're right, it does get worse."

He stopped the car for a minute and then said very slowly, as though he was talking to someone whose English (or IQ) was not up to snuff, "I didn't say it got worse. I said it got *harder.*"

And he was right. When it isn't overgrazed, some very hard country can be excellent cattle country as long as the rancher has the acreage he needs. Someone was running Herefords on the place, and they were in excellent condition. After all, such spare land was the home of the greatest bison herd in the Americas, the very reason that it meant so much to the Buffalo Indians that they would die for it. The Big Dry has a perverse beauty to this day. It is not one vast flat plain like the ones in the eastern Dakotas and Nebraska. It is broken country with an infinite variety of earth tones, the occasional patch of green, and a constant shifting of the light as the land tilts toward the sun in one place and, a few yards farther on, cuts into a shaded ravine that looks almost black against the glare or into a softly lit slope, where exactly the same rocks turn from a pure color to a pastel. It makes the residents appreciate small things. For one example, no matter what rain does to the dirt roads, no one complains. Ranchers in the Big Dry have rain gauges that measure down to the hundredth of an inch (a third of a millimeter).

But before there could be fences or rain gauges or cattle, the Big Dry had to be taken from the buffalo-hunting tribes and turned into private property—a completely incomprehensible notion to a nation of horsemen and tent dwellers. The Felton Angus Ranch on the Tongue River sits squarely in the middle of the geography of the Great Sioux War of 1876–77, the cam-

paign that opened up the country for the Texas trail drivers and
their herds of parti-colored, pied, blotched, and skinny-assed
longhorns. The soldiers of the First Sioux War in 1866 had also
passed through the ranch's land in search of Indians. This was
inevitable. Not only was the Tongue River valley rich in buffalo
in those days, it was also a prime wintering ground for the
Sioux, with two great necessities not available out on the open
plains, water and wood.

When it came time to put an army post on the Yellowstone,
they picked the spot where the Tongue flows into the big river,
a convenient base for contacting or confronting the Indians.
The first commander of the Tongue River Cantonment,
Colonel (later General) Nelson A. Miles, turned out to be the
only successful commander in the Great Sioux War. It was natural that the citizens would name the trading center that grew
up near the cantonment Miles City. It is not ironic that Miles
City is in Custer County, named after the biggest loser in the
history of the United States Cavalry. People were well aware
that Custer, dead, was the prime mover in getting the Buffalo
Indians out of the way.

There are a few anecdotes from the war that help flesh out
the geography and even the mythology that make up the countryside of the Montana and Wyoming plains. Motorists traveling through the Tongue River country or alongside the parallel
Powder River immediately to the east would not suspect they
were following along the routes of the most significant marches
in the Sioux War or passing battlefields and Indian village sites.
Of all the historical trails across the state, the only one thoroughly marked in Montana is an approximation of the route of
Lewis and Clark. It is a little deceptive to put signs along so
many miles of state highway with Lewis and Clark and Sacagawea silhouetted, eternally gazing forward. Lewis and Clark

covered the width of Montana on the Missouri River on the way out, a river virtually invisible from any highway through much of the state. The thousands of Lewis and Clark signposts (which include Sacagawea in the silhouette artwork) seem to be placed as much for local boosting as historical interest. Some of them are fifty miles away from the real route.

The Feltons' Tongue River ranch, which sits squarely on the route of march of the only successful expeditions against the Sioux and their allies, also encloses the site of the last failed diplomacy before the all-out war of ethnic cleansing began. In 1875 special commissioners arrived from Washington, D.C., to persuade the tribes to return to their reservations (and stay

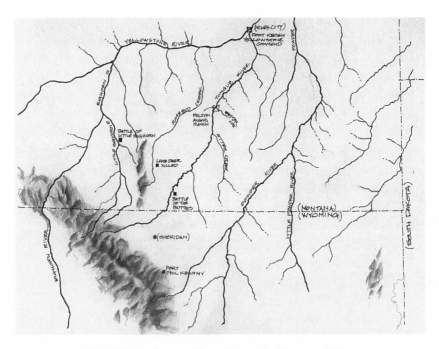

The Felton Angus Ranch on the Tongue River

there). At last they caught up with the largest and also the most powerful group, the two bands of Sioux with Crazy Horse and the medicine chief Sitting Bull. They were at the Sioux's beloved village site on Beaver Creek, where it meets the Tongue at the north end of the ranch. Sitting Bull's refusal to come into the reservation is very likely the only speech ever given on the ranch. (The closest that country ever got to even having a town was briefly having a post office called Brandenburg operated out of a ranch house. The nearest it ever got to being anything but ranch country was in the 1920s when a bootlegger squatted on the banks of the Tongue at the south end of the ranch, grew a little corn, and made some liquor.)

This is how Sitting Bull dismissed the commissioners at Beaver Creek and set loose the dogs of war:

Are you the Great God that made me? Or was it the Great God that made me who sent you? If He asks me to see Him, I will go. But the Big Chief of the white man must come see me. I will not go to the reservation. I have no land to sell. There is plenty of game here for us. We have enough ammunition. We don't want any white man here.

In November 1876, five months after the Little Bighorn, Nelson Miles drove the Sioux with Sitting Bull up into Canada, where Sitting Bull, never short of quotable lines, remarked sourly that "buffalo tasted the same on both sides of the line." Miles then confronted Crazy Horse's band on the Tongue in January 1877 and drove them westward into the open plains. Crazy Horse came in to his reservation in springtime, after a terrible winter of starvation and exposure. Then, in the early summer of 1877, Miles would cross the ranch one more time, running to ground the last free Sioux, Lame Deer's Minniconjou.

Miles's infantry were a strange-looking mob on their winter campaigns, dressed in overcoats of buffalo hide with the hair on the outside, all handmade of each man's devising. Many wore buffalo hats as well, and some had cobbled shoepacs of the same material. Their handmade clothing kept them warm but almost immobile, which makes their incredible marches—including one of nearly a thousand miles to push Sitting Bull into Canada—all the more unimaginable. In the movies it is always the cavalry that chases the Horse Indians, which seems reasonable. In fact, both in the Sioux wars and in the Arizona campaigns against the Apaches, it was all infantry all the time.

The closest thing to a battle on the ranch came in the campaign against Crazy Horse in the winter following the Little Bighorn. A rear guard, staying back with the supply wagons, was stabbed to death just about where the ranch draws water out of the Tongue for the gravity-flow irrigation ditches. The company surgeon, looking at the body of poor Private Batty, remarked how unfair that small battle had been—the dead soldier in such bulky clothing he had barely been able to turn his head or move his arms against a lightly dressed, agile foe. That is not to make fun of the soldiers. Ranchers, bundled and capped and gloved while feeding cattle in winter, are not fashion plates. It is very hard to look like a soldier or a cowboy in a Montana winter. The troops buried Batty on the spot and camouflaged the grave. On their return from the skirmishes with Crazy Horse on the upper Tongue, they dug him up and took him down to the Tongue River Cantonment for burial. This is just as well; what with the ululating screech owls and coyotes, the last thing the Tongue River ranch needs is a ghost out at night.

· · ·

Early adopters of the Aberdeen-Angus breed believed the animals were particularly adapted to the bitter cold of the plains. Indeed, all European cattle are remarkably hardy, inured to both bitter cold and the unfamiliar heat of the North American summer. And of European breeds, it is not surprising that the polled animals (now including the registered breed of Polled Herefords) do well in the coldest of winters.

The inside of a cattle horn has considerable blood supply, and with no insulation at all from the thin sheath, heat dissipates through the walls of the horn. In the dead of winter you can wrap your hand around a cow's horn without fear of frostbite; you will feel the warmth. Studies of large-horned sheep—the physiology of cattle horn appears not to have been studied in detail—indicate that as much as 10 percent of the animal's energy is lost in keeping the horns from freezing. Contrarily, losing heat is a benefit in tropical climates, where enormous horns on cattle are common. The Indian cattle have very large horns (the ones you see in rodeos have been trimmed back to make them safer to ride). The Andalusian large-horned cattle evolved into the classic Texas Longhorn along the hot and humid Gulf Coast. Sub-Saharan Africa also has cattle with huge, long, wide horns. In addition to the benign-neglect, open-range style of ranching that Texans brought so briefly to the High Plains, they also brought a type of cattle that radiated heat into the blizzards.

Cattle's ability to withstand heat is more difficult to explain. Ordinary High Plains summers have spells when week after week the temperature rises over 100° (38° Celsius) every day. This was unheard of in Britain and Northern Europe when the modern breeds were defined and exported to the Americas. It is not that they enjoy the heat; the animals just cope. Truly hot days on the plains change the cows' lifestyle: They drink more, eat less, and seek shade wherever it can be found. The trouble with the High Plains is that the land, by definition, has an insufficiency of trees. Cattle lucky enough to summer in pastures along the Tongue River have a double bonus: They can graze in the shade in the cottonwood groves and, when worse comes to worst, push through a fence to go stand in the river. Cows on the highest paddocks, the ones that rise just a few hundred feet above the valley floor, also have trees—drought-hardy Ponderosa pines that survive on the little extra precipitation that the discontinuous elevation wrings out of the passing winds. But not all paddocks are created equal, and the cattle that lack shade do curious things in the heat that are not seen in temperate or frigid weather.

The hot-day behavior that makes the least sense on first glance is the tendency of the weaned calves, particularly the bull calves, to smush themselves into groups of a dozen or more. Cattle are what biologists call contact animals, that is, they always tolerate and usually enjoy the physical sensation of being in a herd. It looks awful to see a bunch of solid-black heat-absorbing animals crowded together on the kind of day when humans want plenty of personal space and the mere thought of being on a crowded bus or trolley car without air-conditioning brings on malaise.

On one particularly muggy and hot day, a paddock of bull calves on the Felton Ranch were divided into a few huddled masses. Some of the bull calves were crowded in the shade of the only tree in their hillside pasture. It was a skinny box elder, and from a distance it looked like a not very healthy plant in a black pot. Up in the northwest corner there were two more groups, far from any shade, one composed of a dozen and the other of about thirty, and they were packed tighter than grapes in a bunch. From time to time there would be a little jostling, and outsiders got inside and those lying down would rise. It began to make sense. The ones lucky enough to get down first were actually in the shade; their brothers and cousins would stand across their prone bodies. It was a little messy down there, but at least they were out of the sun. With the midday sun not directly overhead (as it never can be at that distance from the equator), animals on the inside of the bunch appeared to be shaded from about the lateral midline down to the ground.

At the latitude of the ranch, the sun never rises more than 64 degrees above the horizon (on June 21), and by the dog days of late August, it is not much higher than halfway to the zenith (53.25 degrees above the horizon, to be precise). Thus, the bull calf standing with another calf between him and the sun is anywhere from two-thirds in the shade to almost entirely in the shade. And best of all, because heat season is also fly season, there were tails swishing around and over all parts of the animals as they stood in random order, both parallel and at right angles, and butt to butt or head to tail. Cows with calves won't bunch as tightly as the single animals, but they will lie down on the shady side of a patch of sagebrush, which, though only two or three feet tall, will give them considerable shade at northern latitudes. Heat does make the animals a bit cranky. They do seem to be a little more stubborn about getting into the squeeze cage for whatever is on tap that day, anything from inoculating them to shaving the hair off their white cryogenically branded identification numbers. The large bulls become extremely cranky on a hot day, if you don't leave them alone. Maurice Felton remarked on that, recounting the experience as a spectator rather than a participant. Maurice does help with the cattle when necessary, but his brother, Richard, does most of the hands-on stock work. In any case, Maurice seemed amused when saying: "Oh, you can move one of those bulls in hot weather. You can move him about a hundred yards. And when he's had enough of that nonsense, he'll whack the crap out of you and go back in the shade." Bulls, being the male royalty on a ranch, are given a paddock with the most shade available.

.   .   .

While Aberdeen-Angus just tolerate heat, they thrive in cold weather. The manner of their lifestyle in Scotland gives little hint to their hardiness. The ancestral black cows in their homeland spent the entire winter indoors, tied to a stanchion with just enough slack to get up and down. On most British farms, confinement indoors is still the case, although group pens have replaced the small stalls of the nineteenth century. If there was one thing that prompted men to perfect the Aberdeen-Angus cow, it was this tolerance for close confinement and stall feeding. It could be done with other breeds, but only the black sleekies had the temperament to thrive indoors. And they were housed tightly in a country much more temperate than Montana.

There are just a few locations in the district of Aberdeen, and none in Angus, with enough elevation that could ever bring temperatures down to routine Great Plains numbers. In a tough winter, daytime high temperatures in eastern Montana hover around zero (–17° Celsius) for weeks on end. Nights at 40° below zero (–40° Celsius) are to be expected. The districts of Aberdeen and Angus are perched on the northeast corner of Scotland facing the North Sea, and miserable cold weather is expected there as well. Up in the Grampian Mountains that mark the western border of the district, there are record lows in both 1885 and 1982 of –7° Fahrenheit (–27° Celsius), but the cows never spend a winter at elevation. In one of the lowest elevations in Aberdeen, near the notable Aberdeen-Angus founding herd at Ballindalloch, the record low is –6° Fahrenheit (–21° Celsius). But these cold snaps in the elevations where the cattle spend the winters are short in duration. Most winters in agricultural Scotland are long, dark, miserable, and wet, but the ground will not stay frozen more than a day or two at a time and frost never penetrates more than a few inches.

It is true that well-fed cattle from any European breed can survive on the High Plains, but not all do as well as the Scotch breeds. As noted earlier, the modern Aberdeen-Angus is a mix of two strains, one small and very hardy that could survive in a crofter's cowyard on the edge of the highlands, and a larger animal suited for the more temperate lowlands. The hardiness genes must be present still.

While traveling in Scotland in May with Bill Sklater, his guests encountered small herds that had just been released from confinement. The animals rose slowly when approached and walked lamely. It would be a few weeks before they recovered their usual gait. Many farmers actually take the cows and bulls, one by one, out on walks, leading the haltered animals for short distances until they get their muscle tone back.

A few farms visited were experimenting with keeping their black cows outdoors in the winter. In both cases, they had some hard stony ground that even when unfrozen would provide a firm footing. At Ballindalloch Castle the herdsman had kept some of the animals outside on the gravelly beaches of the Avon River. A farmer in hillier country had fenced in a ridgetop where the glacial gravel lay exposed on the surface. Bill Sklater was quite surprised at the excellent condition of the outwintered beasts. He professed that he had never seen such a thing and did not expect it.

The main reason that cows are kept in the barn in Scotland has nothing to do with their lack of hardiness. Scotland, like much of Great Britain and the Continent, has very small farms, and every possible inch is cultivated. In the rainy winter season with no permanent frost on the ground, the cattle would tear up the hayfields with their hooves, ruining the dormant grass and clover. On the High Plains, cattle are kept on solid-frozen grasslands or grain or alfalfa fields and fed with spread-out hay brought in by monster tractors. Neither the cattle's hooves nor the well-defined tire treads leave a mark. The other reason for the British practice is the small size of the herds. The average size of a herd of registered Aberdeen-Angus cattle in Great Britain is a little over nine beasts. In 2002 one of the largest farms had twelve young bulls coming on the market out of a total herd of less than forty. In an average year, the medium-sized Felton Ranch will bring 130 to 140 bulls for their annual production sale.

Keeping cattle indoors creates problems for the farmer. The danger from infections is increased, and the mood of the animals can affect their appetite and thus their growth. William M'Combie was adamant about good treatment for the pent-up cows: Herdsmen, he wrote of his employees tending the penned cattle, "must have a taste and a strong liking to cattle—they must be their hobby." He insisted on truly humane, in the sense of individual, treatment: "Cattle feeding in

Haying

the stall should be kept as clean as the hunter or valuable race-horse, and their beds should be carefully shaken up. Many cattlemen are very neglectful of this duty. They ought to take a lesson from the groom in the livery stable. We all know how material it is for our own comfort that our beds be properly shaken; cattle enjoy a similar luxury."

Comfort for the animals brings up the other reason that Scot cattlemen keep their stock indoors in the winter. "I don't know if it's easier on the coos to keep them in," one herdsman said while the farm manager was out of earshot, "but it's greatly easier on me."

Such treatment is possible only on a small scale. The Felton Ranch is a moderate-sized operation by Montana standards, not too big, not too small, just right, and still a hundred times larger than the average herd in Great Britain. The thought of a barn big enough to hold the Felton herd is frightful, and the idea of dragging out the manure of such a host of animals and changing their straw bedding is simply ridiculous. For all the difference in the two systems, each is the easiest way to get the job done in its location and in both sizes of herds. Well, maybe "easiest" is the wrong word. There are very few chores that are easy when the temperature drops below zero (–17° Celsius) on the plains or when intensive labor is required in a damp unheated Scottish barn.

Getting up on a winter morning on the ranch and looking out the living room window at the circular thermometer nailed to the cottonwood tree, one hardly remarks on something as relatively mild as 10° or 20° below (–23° or 28° Celsius). At –40° (it's the same temperature in both systems), someone will remark that it is a little brisk outside. This is, of course, a kind of bravado. Things can get a little dangerous in severe cold. Swinging an ax to break out ice in the watering

Breaking up ice in a watering tank

tank is a clumsy, jacket-encumbered task with numbed fingers inside heavy gloves. One tries very hard not to slip on the ice around the tank and hacks at the ice in the tank with an odd mixture of caution and abandon. A sharp ax might make the job easier, but given the conditions, no one complains about a dull edge.

The cows move with relative caution around the water and the hay feeder in winter, but even in horrid cold a few still have the energy to go bumping heads with their inferiors, maintaining the pasture's butting order. You seldom see a cow fall, although calves, like children in general, seem to regard slipping and sliding as entertainment. What is most noticeable, compared to putting out hay in milder weather, is the urgency cattle show as they get at the dried alfalfa. It is that hay and the water that allow cattle from milder climates not only to survive but to gain weight.

Feeding cattle on the ranch in winter is not enjoyable, but there are some small rewards. There can be beautiful clear mornings with deer and antelope to admire, even if they are picking at the half-ton round hay bales. Ring-necked pheasants get into the cows' feed troughs, pecking for the grains in the cracks and corners that the cows can't lick clean, just their heads poking up until the truck gets too close, and then they cackle and fly. Some mornings a flock of wild turkeys runs alongside the road like windup toys.

Cows have certain advantages over us. Their outer clothing is windproof and permanently water-resistant. In the fall the Angus start growing a winter coat that is not only shaggier on the outside (that varies from animal to animal, some turning almost brown with the new hair by Christmas) but always with an undercoat of extremely fine, soft furlike hair. They are so well insulated that mild-weather snow that melts when it touches the ground will sit, frozen and white, on their backs. In extreme cold, they will be painted with snow, particularly on the upwind side (snow usually comes with wind on the High Plains). In serious snow, the only place it melts on their otherwise insulated bodies is around the nose and inside the ears. As one looks at their heads (they are usually looking right back), there is an ascending *V* of connect-the-dots: a nose, then two eyes, and two ears at the top.

In the worst of winter weather, the cows move into the brushy draws and coulees and groves of cottonwoods, away from the chilling winds. For all the refinement that ten thousand years of domestication have given the Aberdeen-Angus, they have the same instincts for survival as their aurochs ancestors. They are doing just what plains bison would do.

⇥ 8 ⇤

Potential Problems with Cattle Under Domestication

AMERICANS to some extent and Europeans to a much greater degree are concerned about the artificiality of modern foodstuffs. When it comes to beef, the concerns are understandable, given the general fear of secondhand consumption of hormones and the recent outbreak of bovine spongiform encephalopathy (BSE) in Great Britain that killed more than two hundred people and that may yet take more lives, as it can be very slow to develop in humans and all of the victims may not have suffered the symptoms yet. In part by good luck, and especially because of the way Aberdeen-Angus cattle are raised in Great Britain, not a single case of BSE has been found in British Aberdeen-Angus.

As best as it is understood today, BSE infects cattle when they are fed protein supplements that contain neural tissue from an infected animal. The likeliest source, if not the only source in Britain, is feed containing neural tissue from sheep, where the spongiform encephalopathy is so widespread over the years that it has a common name—scrapie. When that understanding was achieved, all nations banned the use of any mammalian protein (and the deliberate or accidental inclusion of neural tissue) in cattle feed. The prohibition seems to have stopped the disease cold in Britain and Europe, although there have been two cases of BSE in the past few years, one in Canada and one in an animal exported from Canada to Washington state.

The source of the disease in these North American cases is unknown, but the animals may have been accidentally, or feloniously, fed mammalian protein feed, which is legal fodder intended only for poultry. Suspiciously, both infected animals had dairy-farm origins. The one discovered in Washington state was a Holstein-Friesian dairy cow; the cow in Canada was a black cow in a herd of commercial Aberdeen-Angus (as opposed to registered stock) that may have been raised, as a calf, on a farm that had both a dairy and a beef operation. To understand how such bizarre things as hormones are introduced into beef cattle and disease-carrying proteins were once routinely fed to some beef cattle and most dairy cattle, we need to look at the perception and the reality of just exactly where we are on the food chain and what it is we might be consuming. Unfortunately, there is a lot of static and noise in the information we receive, and it will take time to explain what has happened and is happening to our meat supply.

One reads in more-or-less respectable publications (major newspapers, national newsmagazines) that the beef animals on the American menu have eaten huge quantities of corn (maize) and other grains. It is alleged that amounts in the several tons (3,000–4,000 kilograms) of grain are required to turn out 1,100 pounds (500 kilograms) of meat on the hoof. *The New York Times Magazine* fuzzed the issue in the summer of 2002 with a front-page headline about a scrawny-looking Holstein steer, announcing that the animal had eaten "several tons of corn and hay." Close reading of the text as it appeared a few months later in book form indicated that the animal had been raised on a corn diet. There are farms in the United States and Canada where cattle headed for the slaughterhouse are given almost nothing but corn for their brief lives. The typical situation is a large dairy where there is no pasture for the milking stock, just a sort of factory floor for feeding them (some of it outdoors but hardly rural, rustic, or bucolic). Neutered bull calves and unwanted heifers of the Holstein-Friesian breed do sometimes get nothing but grain to get them up to weight and out of the way as soon as possible. Frankly, it is a race between fattening the animal and killing it. The diet is unnatural and unhealthy. Holstein-Friesians have been overfed for so many generations that they may have been accidentally selected and have evolved some immunity to an all-grain diet, enough immunity to survive to slaughter weight . . . but not always. The 100-percent corn diet is a program for rearing unwanted stock for the slaughterhouse, not quality stock for breeding or for eating.

If a beef-breed rancher had stuffed a steer with that much grain for the six to twelve months after it was weaned, it would likely kill the animal. Cows (like deer, moose, and elk) evolved to eat a high-volume, low-energy diet. Most of the nutrition from hay (or twigs) is created in the animals' rumen, a kind of auxiliary stomach system where the grass and brush is partially digested by a host of single-celled organisms. Ruminants are really dining on a lot of animalcules and their by-products. The rumen is just a big complicated brewery, turning out food instead of beer. When cows are stuffed with too much high-protein, high-carbohydrate food (as happens routinely on dairy farms), they get very acid stuff in their rumen, which stops the natural digestion and gives them heartburn, or acid reflux. This is not acid they have secreted into their stomach, as in human

Grazing

acidic excess, but a function of the incomplete digestion of food in the rumen, an organ intended to break down poor food and turn it into good food. Dairy cows on a high-energy diet are routinely fed sodium bicarbonate to keep them from serious illness. The other downside of high-energy food in an animal headed for the packing plant is that it not only affects the rumen but changes the contents of the whole gut from alkaline or neutral into acidic stuff. Germs usually found only in the acidic guts of omnivores (like humans) and carnivores can then multiply in the animal's intestines. This is the source of the dangerous strains of *E. coli* and *Listeria* bacteria that then have a chance to get into the meat of the slaughtered animal by fecal or intestinal contamination.

Prevention is simple. If feedlots would put all the animals back on a diet of straw and water for a week just before slaughter, the *E. coli* and other acid-tolerant germs would disappear within a few days. But in the feeding and slaughtering business, no one has a week to spare or clean paddocks to be able to shift the animals into a less-contaminated environment before slaughter. A few specialty packing plants have taken to flushing out the *E. coli* with this simple dose of hay. And beef products sold as "grass-fed," if the label is accurate, are highly unlikely to be infected with any omnivore-carnivore pathogens.

Here is a more realistic example of a beef-bred or hybrid animal headed for a feedlot and onto the packing plant. A hypothetical but typical steer ready for feedlot finishing weighs about 800 pounds (363 kilograms). He has reached that weight, if raised on a typical ranch, without much supplemental grain, and if he's from a ranch with excellent pasture, no grain at all. If the feeder wants at least 70 percent of the 800-pound animals to reach that stage of marbled fat that the USDA calls choice, he will have to add 300 pounds (136 kilograms) to the beasts in about three months. This is the economical best choice, both in time and money. (If someone figures out how to feed a group of same-age, same-size steers to 100 percent choice in the same number of days, he will be a millionaire.) To get steers from 800 pounds to 1,100 pounds (500 kilograms), the weight at which 70 percent of the steers reach choice grade, there are two alternatives. The farmer could just wait another year, keeping the animals (particularly if they are Aberdeen-Angus) on a rich grass-and-hay diet and some supplemental grains. They'll marble up nicely. That's not done much anymore, except for the boutique market share of grass-fed beef. No one has a year to spare for anything. The alternative is to give them a greatly enriched diet for the last few months of their lives. Over that time (and they can eat more and more as they get bigger and bigger), steers can eat and efficiently digest an average of about 22 pounds (10 kilograms) of an enriched ration a day. Typical high-energy feed is 60 percent rolled corn. (Corn is the commonest grain, but wheat and soya may be included in the 60 percent share of the high-energy supplement; the rest is binder, sugars, and roughage.) That will work out to about 13 pounds (almost 6 kilograms) of corn a day for anywhere from ninety to (rarely with decent stock) one hundred twenty days. The goal is to average 3 pounds (1.35 kilograms) of gain a day, but with individual differences, it will take more or fewer than one hundred days to put that 300 extra pounds (136 kilograms) on the steer. The answer is, on average and realistically, a beef animal from a large United States packing plant has eaten about 1,300 pounds (590 kilograms) of corn (or a corn substitute), well short of the "tons" of journalism. These consumption numbers are at the very high end of feedlot practice. Animals can be brought to a marketable state with considerably less food and just a little more time.

Lean Aberdeen-Angus steers grown on ranches like the Feltons' see barely a lick of grain before shipping off to the feedlot. When it comes time to fatten up a few steers for their own use, the ranch animals get maybe as much as 10 pounds of grain a day, just a big bucketful for the three of them poured into a common feeding trough. The ranch animals get it over a somewhat longer period of time than a feedlot steer, and they go through a few tons of hay while they're fattening. This is "less efficient," but who's counting the days? The most recent set of ranch-fed Felton steers (intended for family and friends), when they got to the Miles City packing plant, turned out to be two at the upper end of choice (plenty marbled enough for most people), and the smallest of the three was on the low end of prime. All three would easily have made the cut for Certified Angus Beef, which has to be at the high end of choice and, depending on market demand, may take some at the low end of prime. The three of them, about a ton and a half on the hoof (1,360 kilograms) had eaten less than a ton, about 1,800 pounds (800-plus kilograms) of mixed grains between them. That's 600 pounds (272 kilograms) each, hardly excessive.

The ranch's bull calves, being filled out with an enriched diet in the last few months before the annual production sale, get about the same amount as the ranch-fed steers, considerably less than a feedlot steer. Fattening animals for slaughter is one thing, but it would be unhealthy to make breeding stock that obese. Not that it isn't done. There were two-year-old bulls in the showring at Carlisle, England, in the spring of 2002 that could barely get out of their own way. There is the occasional herd of breeding-stock bulls fed a high-grain diet to their heart's content. A member of the Aberdeen-Angus Society in Scotland, who enjoys taking foreigners around to look at beef herds, watched a bunch of bull calves eating grain and chopped beets and molasses, *ad libitum* as they say in animal science journals. He remarked, loud enough to ensure that the farmer could hear him: "I always like to see the bottom of the feed trough every day." And after we had left that farm, he added that the stockman there had not infrequently gotten back animals that were returned for lack of fertility.

The resulting criticism of beef animals, based on the unrealistic journalistic diet, is that they are much less efficient than poultry in converting food to flesh. The chicken or the turkey is held up as the idolized efficient animal for the carnivores among us. This is a frequently mentioned anti-fact. What follows are averages, but honest ones. Besides not having any rib steaks, the other problem with fowl is that they require a high-energy carbohydrate-and-protein diet from the moment they come out of the shell. A nice 18-pound (8+ kilograms) tom turkey on the butcher's scales has eaten, on average, 98 pounds of grain and protein feed in its twenty or twenty-one weeks of life. In rough figures, it takes 5 pounds (2.25 kilograms) of high-energy grain-based feed to put a pound (0.45 kilograms) of flesh and bone on the Thanksgiving platter. At a ratio of 5 to 1, feed to meat, that's not doing anywhere near as well as the beef animal. Even in its short term in a feedlot, the steer needs a ratio of about 4 to 1, grain to gain. But over a lifetime, starting out on mother's milk and grass and hay and finishing on grain, the ratio is just about 1 to 1. If anyone can figure out how to make a chicken or a turkey grow to market size eating hay and just two or three times its body weight in grain, instead of the normal five times, he will be a billionaire.

The foraged grass and cut hay a steer eats (averaging 20 pounds a day after weaning, and quite a substantial amount from the second through the sixth month before weaning) are hard to calculate as a pure cost. For a High Plains ranch, it does take a lot of irrigated hay for winter feeding and good luck with the weather for range forage. When bad luck comes, the rancher may have to buy hay, the only real way to price a commodity. If a rancher counted up all the hours he's spent irrigating ("irritating" some call it), mowing, baling, stacking, and feeding, and added to that the costs of fuel and spare parts, he could figure out what the price should be, but that would just make him sadder not wiser.

Aside from the putative immorality of eating beef that has taken up too much space on the food-allocation chart, consumers should have some interest in feedlot practices. Beef that is sold as "hormone-free" has probably not spent much time in a conventional feedlot. Steers are routinely treated with hormone implants in feedlots, in the hope that they will put on more weight and do it more quickly and cheaply. The few studies *not* done by hormone-implant manufacturers suggest the cost-benefit ratio is a wash. If your local government's rules for organic labeling include a ban on hormones and antibiotics, that is indirect but persuasive evidence that your beef came from an animal raised sensibly. The sobriquets "grass-fed" and "pasture-raised" should be indicators of safe and conscionable meat, but by themselves they are too vague. Almost all beef is grass-fed at some point and pasture-raised as well.

It should be said that even in the most crass commercial operations, beef production doesn't begin to reach the factory conditions typical of pork and poultry. For instance, cattle are not routinely given antibiotics in their feed, as are poultry and pigs. The antibiotics are not even intended to protect the swine and fowl from disease, contrary to what industry spokesmen will claim, but to kill the ordinary commensal bacteria in their gut. The devastating diseases of factory-farmed poultry and swine are viral, and antibiotics are useless as preventatives or cures. Routine, subclinical bacterial action or "infection" in an animal's gut roughens the walls, causes minute pockets to develop, and slightly reduces that animal's ability to absorb food through the lining of the intestines. The difference in growth rate means but a few days quicker to market for chickens, a week or so for swine. Cattle, on the other hand, absolutely depend on the presence of friendly bacterial in their rumen, that extra set of "stomachs" where microbes digest the cellulose in the fodder.

There is no denying that the grain input in beef in the United States is somewhat excessive, although not at all egregious compared to what chickens and ducks require in grain and animal protein. As for hay, the mainstay of a cow's diet, well, it could be (and is) argued that the irrigated hayfields could be growing vegetarian foodstuffs instead of cattle fodder. But the world where beef cattle live is already overproducing cereals and sugar beets. It is overproducing wheat and beets for the simple reason that there are federal subsidies for some cereal grains and price supports for sugar from any source—cane, beets, or corn.

It should be reiterated that grass-fed beef from one of the British breeds, and the Aberdeen-Angus in particular, does not have to be bright red and without any marbling. Nor does it have to be tough. Naturally raised meat sold in stores that feature organic products is usually as fat-free as nature will allow but not entirely so—any beef animal that survives to maturity will have a reserve of intramuscular fat. The low-fat red beef in the health food supermarket is a marketing choice made by management. It is younger and therefore cheaper than a steer given the time to marble up naturally. And it is more attractive to consumers who fear even small amounts of cholesterol in their diet. As for toughness, the same constraint that creates the unmarbled meat applies. Time is everything, and proper aging (just a matter of weeks) is the clue to tenderness. But as the slogan goes, time is money. Beef that is raised on a healthy diet, held and fed until it is ready for the packing plant without adding hormones or antibiotics, and then aged properly will always be expensive. In fact, it always has been relatively expensive, and it was the possibility of getting a premium price that caused American stockmen to import Britain's most renowned animals to upgrade their herds.

Food writers and consumer advocates and conservationists rightly promote naturally raised beef, as well as the small farms that produce it and the greenmarkets and specialty grocery stores that sell it. However, they all tend to make a common mistake by suggesting that to get old-fashioned natural beef, you need to use old-fashioned cows, rare and uncommon breeds. Old breeds are necessary for producing natural pork and poultry, as the factory farms have developed breeds whose major virtue is the ability to stand not only close confinement but life imprisonment. But it is not necessary to seek out some peculiar and uncommon breed of cattle—Scottish Highland or Durham or Galloway—to get old-fashioned flavor. The great breeds can all be raised outdoors, fattened on grass and hay (given enough time), and brought to market without being routinely fed antibiotics and shot up with hormone supplements. It was exactly those traits of self-sufficiency and carcass quality that brought the standard British breeds to North America. They have been improved but, unlike pigs and poultry, never selected to be imprisoned. A nineteenth-century farmer would not know what to make of a white turkey or chicken, both twentieth-century inventions of the United States Department of Agriculture. The almost hairless and flaccid factory-farm pigs would not please his eye. They would not look as if they could root around in their spare time and find food. But if that time-traveler were to see a twenty-first-century Aberdeen-Angus, Hereford, or Shorthorn, he would feel quite at home. On average, these breeds might look better than they used to—he might think they would die better—but he would recognize them.

·   ·   ·

One of the perils of progress is the almost instant dissemination of a local problem into an international concern. The most notable recent examples are the various infectious diseases that, thanks to airplanes, can leave a location in Africa (West Nile virus) or Asia (sudden acute respiratory syndrome recently, influenza annually) and appear in the Americas without warning. The devastating eruption of hoof-and-mouth disease in England in the 1990s started with a single piece of infected pork flown from Hong Kong and illegally smuggled into the country. And the contrary benefit, of course, is the overnight transport of medical technology and antibiotics. In the world of cattle, the enormous increase in the ease and decrease in the expense of moving cattle by ship and train, by truck and plane, did allow for the improvement of stock everywhere, and also inevitably the endangerment of animals worldwide. Besides the issues of infectious diseases, which are clear in their symptoms and for which a combination of medical intervention, elimination of infected animals, and quarantines are successful treatments, the ease of transportation in the twentieth century led to some more subtle plagues.

Genetic diseases, unlike fulminating infections, can lie hidden for years. It may be generations before the disease expresses itself, and by then it may have spread throughout the country. It was the very ease of transportation that allowed genetic diseases to go unchecked in the United States for decades, but only after the diseases had appeared spontaneously and very locally, indeed in a single individual. The genetic diseases that were spread by modern transportation went unnoticed because the herds in which the genetic flaw occurred were not managed in the ordinary manner.

The Felton Angus Ranch, like most breeders of registered stock, uses a combination of natural and artificial insemination. Natural breeding, with your own bulls and cows, has its benefits. Good traits from animals the stockman knows intimately are almost guaranteed, and it's much the simplest way. Decisions do have to be made about which cows are going to breed with which bulls. Artificial breeding using purchased sperm has the benefit of introducing new bloodlines into the herd and perhaps shifting the general physique of the herd in one direction or another. And it has the advantage of avoiding too much inbreeding or, more precisely, too much line-breeding. But avoiding line-breeding can be taken too far.

A rancher using his own stock, breeding cousins and siblings and fathers and daughters, does get one important benefit. If, heaven help him, there's an inheritable problem, particularly a problem caused by one or more recessive genes, the only way to find it before it gets out into the marketplace is by repeatedly crossing close relatives. Breeders seldom encounter serious genetic problems, but when they happen, they can be devastating, destroying livelihoods, bankrupting corporate farms. One of the worst of all the genetic defects of cattle produces dwarfed animals. This has happened twice in the twentieth century. One outbreak was within the registered Angus industry, but the first and most widespread eruption of dwarfism happened with greater frequency and much greater economic destruction in the other major British beef breed, the Herefords of North America. Although, as stated earlier in the chapter on the evolution of the domestic cow, nonlethal miniaturization is a possibility in the genus *Bos*, it is by no means the most common form.

Dwarfism is the most frequently seen kind of miniaturization in cattle, and it produces calves that are unlikely to survive to adulthood. They have not only short but distorted legs. Often they are swaybacked, and the most characteristic deformity is in the head, with dwarfed nasal and frontal bones that give them a dished face. They have difficulty breathing and from this debility get their colloquial name: snorters. In nature or in feral herds of "wild" cattle, a mutation producing snorters would be quickly reduced to a minimum, and if the herd in which it appeared was closely inbred, it would probably disappear entirely by auto-destruction. But in the more benign world of raising breeding cattle, where animals are kept from much inbreeding and where offspring are disseminated widely, the natural control breaks down.

In the century just past, a severe epidemic of snorter calves broke out in North American Herefords, with the first cases appearing just before World War II, and increased geometrically continuing well into the 1960s. The reason it persisted so unnecessarily long was a human problem: unwillingness to face the truth. Although no one had ever seen a gene or even the much larger spiral helix of chromosomes, when the epidemic of dwarfing began, the practical genetics of animal breeding were widely understood. Most important, the idea of recessive genes, which require a copy from both parents to produce the effect, was thoroughly understood. But for more than two decades the Hereford industry refused to admit the possibility that there was a rogue recessive trait spread in varying degrees throughout much of the registered breeding stock. For years Hereford blamed it on diet, on mineral deficiencies or excesses, on aberrant insulin uptake, on random variations in blood chemistry, on weather. The animal husbandry journals of the 1940s, 1950s, and early 1960s are replete with accounts of very careful scientific pursuit of each of these dead ends. The closest the Hereford breeders got to admitting it was genetic was blaming it on a few breeders who had developed Herefords with extremely short legs and deep bodies. It was not their fault. When those bloodlines were examined carefully, it turned out that they were actually less infected with dwarfism than the standard type.

As the truth began to sink in—that there was no possible explanation except a pervasive recessive gene that was particularly common among some of the most prominent herds in the United States—there was a short-lived attempt to cover it up. The Hereford Association went so far in the winter of 1954–55 as to forbid its employees to publish, review, comment on, or even speak privately on the question of which bloodlines were carrying the gene. This prohibition was at the behest of a collection of fifteen large breeders of dwarfism-compromised registered stock who were starting to lose serious money. This cabal, by the way, named itself the Committee for Hereford Progress. At the next convention, in the fall of 1955, the largest number of members of the association that had ever attended voted out the officers of the association, replaced them with a reform slate, and also voted to disband the Committee for Hereford Progress, all by a margin of nine to one. That done, and with painstaking retrospective research, tracing back the bloodlines began. It took years. At that time, almost no one outside the federal government had computers. The IBM punch card was regarded as the height of data storage and retrieval. The Hereford registrations were on file cards or written in ledgers. Sometimes the accumulated registrations and herd books of breeding stock look like exercises in self-promotion, but when a widespread genetic problem surfaces, they are the only chance at running down the source. The crucial work was by a single dogged researcher from Iowa State, L. P. McCann.

Hereford dwarfism is one of the great cautionary tales about excessive dependence on outcrossing or pathological avoidance of line-breeding. In the first year of the twentieth century, the owner and the hands-on manager of the WHR ranch acquired a year-old bull of impressive breeding. This was Prince Domino, born in the 1890s. He carried the mutation of a recessive gene that could, when both adults had it, produce snorter calves.

It would have shown up fairly quickly, as soon as children or grandchildren or first cousins were bred in numbers. Like most recessive genes passed when both parents have a copy, on average one-quarter of the offspring will be free of the defective gene, one-half will carry it, and the remaining one-quarter will express it. These are just averages, and that is why it may take several crosses between closely related animals in a herd before the breeder can either locate the source of a problem or be sure that his herd is free of it. When only one parent has it, as in the Prince Domino case, on average half of the offspring will carry just the one copy of the gene and the other half will be free of it entirely. The trouble starts when the descendants carrying a single copy breed with one another.

The herd manager at Prince Domino's Wyoming ranch had started in the cattle business managing a Holstein-Friesian dairy herd near Denver, Colorado. There, he had engaged in considerable line-breeding. It was understandable; artificial insemination was sixty years in the future and Holstein bulls, as much as any dairy bulls, must be treated as potential assassins. One is all you want, and you tend to stick with the devil you know rather than importing trouble. The bull was bred back, time and again, to daughters and granddaughters, as were his sons. But the close line-breeding eventually produced a host of genetic problems that effectively ruined the herd. The manager swore never to line-breed again. Not ever. Not at all.

The result was that the recessive gene for snorter calves in his Herefords was disseminated widely before, by accident, two distant descendants of Prince Domino, many generations removed, bred together. The common and sensible practice of regularly bringing in new lines for breeding began to make the connections between some of the many widely dispersed descendants of Prince Domino, ones who carried the defect. It was effectively scattered all about the country, perfectly hidden, waiting for the happenstance (but inevitable) breeding of a bull and a cow who each descended from Prince Domino, and who each carried a single copy of the recessive gene. Weeding out these carriers ruined several breeding ranches, and hurt even stockmen whose cattle were free of the defect because they lost a good deal of market share. Fear of buying a carrier encouraged breeders to look overseas for prime Hereford breeding stock that they could feel confident about, thereby depressing prices in the United States.

As night follows day, the dwarfism curse moved on and in a few years, in the late 1960s, began to appear in North American Aberdeen-Angus herds. Again, it had sprung up spontaneously in just a few animals, but because of deliberate avoidance of inbreeding in those herds, it was not noticed until the distant relatives, generations after the gene appeared, began to meet as other herdsmen brought in "fresh blood." Given that the Hereford people had just been through the same event, it might be expected that the Aberdeen-Angus breeders in the United States would start searching back in the genealogy of the parents of dwarfs and locate the type and get it out of circulation. But no, they essentially repeated the same acts of denial as the Hereford breeders. They looked for the same nongenetic causes that the Hereford breeders had pursued, and like the Hereford herdsmen, some began to blame the Aberdeen-Angus breeders who were turning out small stumpy animals for creating the pathological dwarfism.

The fight within the Aberdeen-Angus Association of the United States did not get as ugly as the recent fracas at the Hereford Association, although there was one serious and devious attempt to avoid the truth. A breeder who had paid an enormous price for the time, $60,000 for a bull and $35,000 for a cow from a herd that was riddled with carriers of recessive dwarfism, tried to protect his investment by forming a bogus Committee of Angus Breeders that asked the association to stop collecting information about and pedigrees of dwarf-gene carriers and to order the association staff not to discuss dwarfism in public or private. The source of the letter was easily identified (there was no committee), and a few months later his herd manager wrote the association and asked that the matter be dropped. The economic costs to breeders who had accumulated the dwarfism gene in their herds were enormous, and as happened with the Hereford breeders, although the problem was not as widespread, some were bankrupted out of the business.

Breeders of the Aberdeen-Angus cattle also followed the lead of the Hereford breeders after the disease was understood and looked abroad for safe investments. Dwarfism was unknown in British registered herds, and a goodly amount of expensive British breeding stock was imported. Ironically, considering that pathological dwarfism was the motivation, at the moment in time that Aberdeen-Angus breeders in North America looked toward Britain for stock, farmers there were deliberately producing small but normal animals. Because that fad coincided with the renewed North American interest in British bloodlines, small bulls became fashionable in the United States, and the average size of purebred Aberdeen-Angus steers and heifers going to packing plants decreased as much as 30 to 40 percent in North America as it already had in Britain.

Outbreaks of genetically caused diseases are extremely rare in cattle compared to the many recurring genetic problems that bedevil companion domesticates like dogs. There is not and never has been a cattle phenomenon as widespread as the problem with hip dysplasia in large dogs, particularly in strains of Labrador retrievers, nor of the various eye and nervous system diseases that occur in pet animals. Cattle are bred for utility and, for the most part, bred by practical men. Very often one hears a dog owner with a badly limping Labrador remark matter-of-factly, "Oh, it's genetic, you know," or the same explanation is heard for the cataract-blinded cocker spaniel. That kind of thinking isn't tolerated in the world of cows. Cattlemen do not expect flaws, and they do not tolerate finding them in breeding stock they purchase. As the process of decoding the genome of domestic cattle proceeds (it is close to completion), inherited genetic defects will practically disappear because genetic testing will be possible, just as it has become possible for human diseases like Tay-Sachs disease or Huntington's chorea. But in the meantime cattlemen (and dairymen) who stick with established breeders of integrity and established bloodlines of equal integrity have little to fear, wherever they look for breeding stock, semen, or embryos for transplantation.

What brings customers back time and again to the registered herd at the ranch of their choice is an odd mixture of motives. Certainly the first one is faith in the breeder's honesty and attention to detail, including assurances that the herd has no genetic time bombs. With some qualifications (all guarantees have them), registered beef cattle breeding stock is guaranteed free of genetic defects, although the only recompense is the purchase price of the animal. And a second motive is that while ranchers may want more bulls like the ones bought before, they also hope that the next batch will be a little different, better, and carrying a new mixture of strains. Cattlemen do believe in progress and, to a point, do believe that newer is better. When trust in a herd's genetic integrity is established, buyers may even purchase animals sight unseen and without knowing the precise lineage.

As one example, and it happens everywhere, an absentee bidder wanted to be represented at a Felton production sale. He was tied down at his home ranch in Kansas and couldn't make it up in person. That led to a succinct conversation between Richard Felton and a representative of the American Angus Association, who would stand in for the buyer. Society officials make good surrogates, being experts on the animals and also on their cash value. The buyer sets a dollar limit and presumably gets his money's worth with the price set by the market, not the breeder. It is one of the few examples in ranching where pure market economics are in play.

This is what Richard explained to the bidder: "This fellow

The Felton production sale barn

wants a bull with good feet and a long body, and a deep body, a high weaning weight, and low birth weight." After a pause, and with no tone of irony, he added: "And for $3,000 or less. I guess he wants the perfect animal." What the man wanted was a maximum of good traits, but first on the list was the rather curious "good feet." As you will see, it will make sense after looking at the other qualities the Kansan desired.

Low-birth-weight calves make for easy unassisted delivery, and a high weaning weight ensures that the smaller calf is going to get up to good market size in as short a time as possible. The long body is everyone's desire. The high-priced steaks, rib eyes and sirloins and tenderloins, come out of the latitudinal direction of the animal. Cheaper cuts, round steaks and rump steaks and hamburger and stew chunks, come out of the vertical parts. Long-legged steers are good for burgers, not so good for steak houses. Deep bodies just enhance the value of long bodies—more pounds of rib roast and rib steaks and that odd, dry but highly prized cut, the skirt steak.

Like everything else in cattle breeding, it becomes a trade-off. For a time, first in Britain and then in America, breeders turned out bulls and cows with such a deep body and such short legs that they showed little daylight between their bellies and the ground. That might be all right in a demonstration barn, but it just doesn't work out on the range. A cow's got to walk, and bulls amble miles a day in breeding season, moving from cow to cow, looking for one open and breedable. The energy used up walking is proportional to the length of the legs. Longer legs mean longer strides without any more effort expended. So, to breed an animal that can thrive and reproduce on the open range and in pastures measured by the square mile, you have to leave enough leg (low-value leg from the butcher's point of view) for the cows to find enough to eat and the bulls to find the cows in heat.

One of the consequences of too much breeding to increase the size of the animal is that the mature cows, and particularly the enormous mature bulls, can outgrow their feet. By putting feet at the head of the list, the out-of-town buyer indicated very clearly that he did not want a grotesque animal—he wanted a range bull. There is apparently more plasticity in total size than in hoof size. (This makes common sense. There are so many dimensions to alter genetically that add up to total body size and body conformation. The genetic palette for the hoof is much narrower.) Cattle have cloven hooves, a design found also in swine, goats, sheep, and all the members of the deer family. It makes for a foot that provides decent balance and some conformity to uneven surfaces, but it also creates more perimeter to crack. Also, the cleft between the toes can be the focus for a major problem—a growth, a split, a dislocation of the toe bones. So it was that the optimistic (and trusting) absentee rancher wanted, first and foremost, good feet. If it was possible, after meeting all of those demands, the rancher who stayed home in Kansas had added that he would like a bull with a pedigree that included a good dose of genes from a certain Scotch Cap. The Feltons have lines with lower birth weights and higher weaning weights, but Scotch Cap is a good middle-of-the road sire.

Richard makes no point of it—it's not included in the sale catalog notes—but the Felton herd has had considerable though intermittent examples of deliberate inbreeding. By occasionally bringing closely related animals together, he gets reassurance that his herd hasn't acquired some negative recessive traits. But for about three-quarters of a year's production of calves, their conception will be man-managed, using new genes from outside the ranch. Artificial insemination (AI) is a kind of "keeping up with the times" procedure. By trying different lines, and without the expense of purchasing numerous bulls, breeders stay current with improvements in the breed, stay on course toward the unattainable goal, the perfect animal.

Compared to buying and using a new bull, upgrading by creating a crop of calves with AI is time-consuming and requires a certain sensitivity in the fingers that is often not compatible with a lifetime spent mending fence and roping calves. For most dairy herds and many ranches, specialists are brought in, sometimes a veterinarian, more often men and women with no other role in life than wandering from dairy to dairy and herd to herd, making calves the new-fashioned way. Richard Felton has both the talent and the temperament for artificial insemination, a procedure requiring equal parts of manual dexterity, patience, and perseverance.

→ 9 ←

Man Makes Cow

A BRIEF description of the artificial insemination process follows immediately, and then some attention is paid to the usual mode. On a ranch, artificial insemination always precedes natural insemination. On a registered-stock ranch, the common practice is to breed a large proportion of heifers and cows to outside bloodlines, using previously frozen semen. After the AI process, ranchers will turn all the females out in different paddocks, each with a bull. If the AI didn't take, the ranch bull will breed the not-pregnant ("open") females. On typical extensive commercial ranches, that is, places where crossbred cattle and unregistered breeds are raised for meat, all the work is done by bulls, occasionally borrowed animals but usually owned by the ranch. Pure artificial-insemination-only beef herds are rare in North America, usually very small farms without sufficient room to keep a bull or research herds at colleges and field stations that are continually mucking about in the gene pool and couldn't house enough bulls to satisfy their experimental needs. As noted earlier when discussing cattle behavior, most dairy herds are entirely articially inseminated, in part for reasons of safety.

If AI could be started the very hour that a cow comes into season, it would be a simple process. The rancher or technician would just take a short syringe (the mythical turkey baster, for example), thaw out the frozen sperm (they've been hibernating in liquid nitrogen), and squeeze the sperm into the cow's vagina. The tiny half-an-animal sperm cells would wiggle their tails and swim on up, pass through the cervix into the womb, take a left or a right into the uterine horns (feminist jokes about the mentality of sperm aside, they have no apparent clue as to which horn is involved in the particular cycle and randomly go one way or the other), meet the descending egg, and one of them would penetrate the egg and begin the process of pregnancy. If both horns have an egg, fraternal twins will result. This is what bulls do naturally. They sense the pheromones that mark the beginning of ovulation, and if the cow, who has her own sense (her own hormonal clues, not common sense), agrees that it is time and will stand for it, he'll mount immediately and inseminate swiftly. To have good timing requires constant vigilance, and the bull has to cover a lot of territory and smell a lot of cows.

Unfortunately, human noses (or the olfactory nerveprocessing centers in our brains) are too crude to allow stockmen to imitate nature. Ranchers must wait for more obvious signs of ovulation. Fortunately, from the human point of view, cows and calves and bulls all respond to the pheromones of ovulation as if they were capable of inseminating their pasturemate. Cows and heifers waiting to be AI-ed will mount the ovulating females in their herd. This is called bulling, and all cattle do it to each other regularly and spontaneously. The difference with cows coming into breeding season is the frequency and intensity with which they mount each other. The act is common then, but otherwise only occasional.

Although it usually has nothing to do with female phero-mones, young bulls are particularly prone to mounting each other. Their main recreation is reinforcing the butting order, fol-lowed by bulling. If they smell a cow in heat, even at a consider-able distance, this will really put them in the mood, and lacking any likelier partners, they will mount each other promiscuously. Whether bulling, like head butting, is a show of dominance is a good question, and the answer is probably "yes." Toning down their dominance behaviors is one of the main reasons for cas-trating bulls and turning them into steers, and while a pen full of steers will occasionally have a bulling incident, it is not nearly as frequent as with same-age bulls. A feedlot full of bulls would not be a happy place to manage. As it is, the occasional steer that constantly wants to mount will be pulled out of the fattening pen and sent right on his way to the knacker. As many as 5 percent of the steers in a crowded feedlot will be seriously in-jured while bulling and being bulled, even with vigilance on the feeder's part.

The difference in the corral full of cows waiting to be bred is that the ovulating cow is not only the center of attention, but when she is just coming into heat, she will put up with it. Some yet-unspecified hormone of the several released into the cow's brain at the time of ovulation alters her mood, and she will allow herself to be mounted. At this time, as ranchers say with classic western brevity, the cow "is standing." There are other hints. The ovulating cow may bump heads with her herd-mates more frequently; she will pace nervously (cows ordinarily waste no energy). In warm or humid weather, she will sweat visibly (thus, "coming into heat" is descriptive). Her external labia may swell, but as they are usually covered by the dangling tail this is not a clue easily observed. A cow that has been mounted while no one was watching has fairly obvious clues; a wet back and mussed-up hair is almost diagnostic. And a good AI-er just knows—it is an art. But however skilled the AI-ers are, they have necessarily missed the magic hour just before the egg begins to descend into the uterus, the moment when the sperm still had time to whip-kick all the way upstream. Since the AI-er is late, having waited for more obvious clues that become visible only very close to or at the moment of ovulation, he has to cover most of the distance for the sperm so it will catch up with the descending egg while the ovum is still in the uterus.

Some ranchers and dairymen don't like to spend time searching out the cows in heat themselves, so they use an animal that has been altered to do the work. An old dry cow may be given a large dose of testosterone by injection, and this will turn her into a bull, as far as behavior is concerned. She will find the ovulating animal with as much precision and interest as the real male would. The other choice is to get a bull, and either castrate him and kick up his hormone level with injections or do unpleasant things to his penis, preventing him from penetrating and inseminating. These altered beasts are called gomer bulls.

Richard Felton sorts the heifers and cows himself, and it is one of the more pleasant tasks, at least for a visitor. The candidates for AI are corraled up together, and each morning and afternoon (into the evening when the days are long enough), he walks through the corral, looking for any and all of the signs of heat. Walking among the animals is another way to understand the distinction between a "domesticated" and a "tame," or "pet," animal. It is a rare cow that comes over to be caressed, although that can happen and would happen more often with fewer beasts and more personal attention. That would be a pet cow. The Felton Angus are docile, not hostile, but they are not that tame—they are very wary. A stranger in the Felton corral is in little danger. The cows and heifers move away immediately and, unless cornered, won't come near a visitor. Richard can get closer because he is a familiar.

Strangers can help sort cattle after finding the ones in heat, as long as they understand the rules of cow behavior and use this intense avoidance reaction purposefully. With one gate open in the big corral, leading into a corridor to a smaller holding pen, one or two people can move the AI candidates just by using the cows' instinctive desire to keep away from humans. They will move away as you approach their personal space. More often than not, they will move in the same manner as the human being. Around slow and calm people, cows move slowly and calmly as a rule. There are always some hard cases. It is something like that endlessly entertaining childhood activity, pushing one magnet around with another one aligned positive to positive and/or negative to negative. Sorted, they will stay there until early afternoon, and then be AI-ed. If sorted in the afternoon, they stay overnight and are treated in the morning. It takes longer for an amateur to sort the cows, not because they won't move but because they are very wary, and more of them will head off for the sorted-out pen than the one or two wanted ones. The easiest thing to do is let the small mob head for the sorted-out pen. If they insist on going inside in a group, it's a good idea to let them. There is no use arguing with a cow. Once they've settled down in the small pen, it is much easier to use their avoidance to move the cows in heat away from the gate and let the ones not wanted slip by and out into the corridor and back to the main corral.

It can be pleasant work. Even while tromping around a mixture of frozen and fresh cow flop (an extremely uneven and slippery surface) in a Montana January, there is some pleasure in being so close to the animals. This may sound odd, that one would enjoy being nearly a member of the tribal herd, but it is part of being a rancher, a herdsman, a farmer. In Scotland they call this sorting process shedding cattle, because their pens are always roofed-over. In his day William M'Combie of Tillyfour in Aberdeen was noted for the ease and efficiency with which he shedded his own cattle. It was a high compliment and he did not have to rely on their avoidance reactions. His cattle were quite fond of him, and one visitor to Tillyfour remarked on how they would come up and sniff his pockets, hoping for a treat, and almost as a joke, one of the cows would mouth M'Combie's handkerchief out of his jacket pocket.

When the animals have been sorted and the time for artificial insemination has come, it gets more complicated. The cow's cervix (literally, the "neck" in bio-Latin), which separates the uterus from the vagina, is small, and in heifers (virgins, so to speak) it is tiny. The only way to thread the sperm syringe through the cervix into the womb, to get the sperm to the ova on time, is to palpate the internal organs and locate the cervix. We are all, male and female, aware of the rectal examination of parts of our reproductive anatomy. It is the same with cattle.

The cuff of the AI examination glove is all the way up to the armpit, not the wrist, as it is for humans. The

Artificial insemination

cow's lower intestine is very large in diameter and very flexible in its walls. If right-handed, the AI-er pushes his left arm up the rectum to his elbow at least, grasps the bottom of the intestine between the thumb and the first two fingers, completely encircling the cow's uterus, and slides his hand back and forth and feels for the O-ring of the cervix. Unlike the soft and squishy walls of the vagina and uterus, the cervix is firm. It has been described as feeling like a chicken's neck, but, inconveniently, it is considerably smaller in diameter; a pigeon neck would be more like it. In heifers, the cervix is no larger around than a bean; in postpartum cows it may be as thick as a smallish unshelled pecan. Most ranchers, even if they do some AI themselves, let bulls or professionals handle the heifers, most ranchers having neither the temperament nor Richard's talent for AI-ing adolescent cows.

Since the insides of cows do slop around, there is enough slack in the intestine to let one grasp the cervix. With the cervix located and grasped to hold it firmly, the sperm gun is inserted carefully at an upward angle to make sure that the slender sperm-shooter doesn't accidentally enter the cow's ureter, which opens into the bottom of the vagina some four inches from the entrance. The gun is a clever tool in three parts. There is a hollow metal tube with a plunger inside it. The packet of sperm, thawed out after being removed from the liquid nitrogen tank, slips into the open end of the metal tube, and the packet's protruding end is snipped off neatly. Then the third part of the gun, a plastic jacket that fits over the metal tube, is slipped on. This plastic jacket has one completely open end that goes over the metal tube; the other, partially open end has a lip around its inside edge that catches the rim of the semen tube when the plastic device is pulled over the metal tube. Thus, the outer tube holds the sperm in place until the interior rod, the plunger, pushes forward, ejecting the sperm. Although rigid, the sperm gun works on the same principle as the long and flexible shutter release used on cameras held steady on a tripod. As pushing in the handle of the shutter button moves a cable that triggers the camera's shutter, so pushing on the sperm gun's rod ejects the sperm.

When the cervix is in its natural position, it extends out and rearward into the vagina so that there is a surrounding concavity (the *fornix vaginae*) where the sperm gun can easily hit a dead end. If the reader wants to visualize this, imagine a partially inflated rubber balloon as the vagina and the inflation tube as the cervix, and now push the tube a half inch (12 millimeters) into the balloon so that, seen from the interior of the balloon, the tied and rigid inflation tube is surrounded by a circular concavity, a moat of tissue. Since the interior diameter of the cervix is not much greater than the outside diameter of the sperm gun, it may take several seconds of probing before the gun passes through the opening. Just to make it particularly complicated, a cow's cervix is not a simple ring. It has three successive ridges, three accordionlike pleats, and the tip of the sperm gun can hang up on any of the pleats. The way to avoid catching up on the ridges or in the *fornix vaginae* is to grasp the cervix (but not too tightly) and push it forward toward the cow's head. Pushing the cervix away from the vagina reduces the depth of the fornix surrounding the cervix. It also elongates and stretches out the cervix, as when an accordion player moves his hands apart and unfolds the squeeze box, thus pulling the pleats out of the way. As the AI-er holds and feels, the next step is to draw the tip of the sperm gun back to the upstream, forward, end of the cervix and bung in the sperm. The sperm must be released close to the cervix so that the sperm have an opportunity to paddle their way up one or the other of the two horns of the uterus so that the one with the egg is fertilized. The horns branch off almost immediately upstream of the cervix.

Most cows are not amused by the procedure, although they do not seem in pain. Confinement in the squeeze cage annoys many cattle no matter what's happening to them. The restless ones hate the cage itself, probably realizing something nasty is up (just as some dogs resist going to the vet's office). Even if all that's scheduled is an examination, they'll fight the cage. Really obnoxious cows will go down on their front knees, putting their internal organs at an acute angle downward toward their head, which just makes insemination more difficult. What they're really trying to do is lie down flat, in a defensive position like a hiker attacked by a grizzly bear. But the narrowness of the cage and a bar behind the hind legs (meant to keep cows from kicking their inseminator) makes lying down impossible. Because the uterus is now downhill from the animal's rear end, it may be necessary to clamber up the insides of the squeeze cage, balancing with difficulty, in order to probe down into the kneeling cow at the proper angle. Watching someone try to inseminate an ornery cow explains why there is a whole occupational industry of hired sperm-guns.

The average annoyed cow will just slam around, which is hard on the left arm since it's quite a ways up in there, and it also causes considerable pain in the shoulder at the rotator cuff. A cow can also rock back and forth the full length of its neck. The pinching bars on the cage don't squeeze the cow's neck itself; they are just tight enough that the head and shoulders can't get through the opening. That's more than enough range of motion to knock down an AI-er who can't respond quickly. Add that he or she is standing on a metal floor covered with fresh cow excreta, which is trying to freeze solid in normal plains winter weather, and you can see that AI-ing is part clinical and part acrobatical.

British animal husbandry texts encourage the practice of massaging the animal's labia and clitoris in order to relax the cervical opening and perhaps change hormonal conditions inside the uterus to increase the chance of fertilization. Whether stimulating the cow actually improves the odds of reproduction is unknown. American cowboys do not fool around with cows, and mention of this British practice brings sniffs of derision. They do make contact with the cow's external sexual equipment because the rectum is above the vagina, and there is usually some manure that needs to be wiped off before inserting the sperm gun. With one particularly spasmodic and unruly cow, Richard kept wiping away with a piece of paper towel until the cow stood still. "Does seem to calm her down," he said, without smiling.

A healthy herd requires bringing in new genes from time to time, and AI means you can bring them into Montana from Arkansas or Alberta and ship out genes to Argentina. Frozen semen has no real capacity to convey disease and there is usually no quarantine. And it's cheaper not to airfreight the whole bull. It costs very little to ship enough sperm to impregnate a thousand cows. It costs a great deal more to freight the living bull even a few hundred miles from Canada to the United States, but then quarantines would be required, and the rancher would need two or three bulls to cover just a hundred cows. AI also allows the rancher to increase the productive life of his own best bulls. Long after a bull can cover a pasture full of cows, after old age and attendant arthritis and bad feet have slowed him down to an amble, his collected and frozen and thawed and injected sperm lives and reproduces. One of the ranch's best bulls, FAR Iditarod, named after the sled-dog race in Alaska, was still producing offspring three years after his death. He turned out smaller-than-average calves that grew much faster than average. Until the supply is used up, Iditarod's excellent genes will work away in breeding season.

While cows and cowgirls (as cowboys are much too fond of reminding them) share a monthly cycle and a nine-month gestation, it is much easier for cows to get pregnant. The cow egg, sliding down the wall of the uterus, just needs to hang on to the wall to get fertilized and become a healthy embryo. For a variety of reasons, most especially the primate system of blood supply to the fetus, the human egg has to effectively burrow into the wall to make it through the first few weeks of pregnancy. But easy as it is with cattle, and as careful and skillful as Richard is at groping up the rectum to feel the AI syringe pass through the cervix, success is seldom more than 70 percent per attempt. With hundreds of cows to impregnate, it just takes too much time to get up to anything approaching 100 percent pregnancy by AI. The odds don't improve as time goes by, and perfection is impossible. Re-inseminating twenty-five cows out of a hundred that don't catch the first time gets you another eighteen pregnancies at best, assuming that all of them are capable of becoming pregnant. It would take a third session to push the success rate over 95 percent. That is about as high as the percentage goes, as much success as the rancher can expect, however the cows are impregnated. It is Zeno's paradox in action—the arrow that never reaches the target, the fox that never catches the rabbit.

Some efficiency in the labor of AI (but not in the success rate) is achieved by forcing the cows into a state of ovulatory heat with hormone injections. Ranchers use different combinations of oral medicines, implants, and injections, all with the same purpose: to get all the beasts coming into heat within a few days of one another. This lets them AI for a few days and then get back to work on everything else that's been put off. It also concentrates the calving season at the time you want and gets that over with more quickly. Stretching out the calving season might seem as if it would spread out the work and make it easier, but cows going into labor have to be checked on from dawn to dark, and heifers, who often have difficulty delivering their first calf, have to be checked at intervals all night as well.

Repeated AI is feasible in a research herd or in a small specialized farming operation, but that means doing internal physical exams of each cow for signs of pregnancy. It takes way too much time for a working rancher. Identifying the still-barren ones (the ranch word is "open" for unpregnant animals) and injecting them with hormones requires a second trip to the squeeze cage. Waiting and getting them back into the squeeze cage a third time for AI a few days later (by which time they are in no mood at all to put up with it) is a chore avoided and a problem most practically solved by the use of the actual bull to cover the cows not yet in the family way.

The whole group of cows, both pregnant and open—it's not worth the time and trouble to figure out who's done and who isn't—goes to the pasture with the bull. It will be a few weeks before the open animals are ready to breed, and just to make sure the bull has covered the herd as best as possible, he is left with the cows for another thirty-odd days. Determining whether a cow was inseminated by the bull or artificially by Richard is simple. Cows deliver within a few days either side of the date that marks the end of a nine-month pregnancy. Heifer calves tend to come in the early part of the cycle, bulls somewhat later, but there is considerable overlap. The dates of birth will confirm the paternity. Nine months after AI means the mail-order groom succeeded; calves born four or eight weeks later belong to the bull. If there's any question, say a particularly handsome calf that is going to be sold as breeding stock that has an intermediate birth date, DNA testing comparing the calf to both possible sires for about twenty-five dollars solves the problem and gives an accurate pedigree. One cause of an intermediate date—a rare occurrence—is that the AI-ed cow had not come into heat before the procedure and then ovulates soon after going out to pasture with the bull.

As one reads the newspapers or watches the television, it may seem that everything in the world is done more efficiently or more excitingly if it is done artificially. Artificial insemination has become a widespread practice for the simple reason that it is cost-efficient for the specialized breeder or the dairyman. An even more complicated form, the embryo transplant, can bring a new line of female genes into a herd as well as the new male genes so easily transferred by AI. Embryo transplants are much easier in cattle than humans, again as with the internally fertilized egg, because the process of attaching to the uterus and gaining a blood supply is less complicated in cows than in humans. Cloning, which is very fancy AI when you get to the bottom line, is being promoted as the new great method of animal reproduction, wave of the future, etc. Dolly is the only sheep in the world to ever become famous. The trouble with cloning is that it impedes improvement. It may be useful for something unnecessary but still desirable, such as creating two identical horses to pull a carriage, but it doesn't move a program forward. Willie McLaren sells his brood cows when they're still producing calves, because, as he said, "If I'm doing it right, my heifers should be better than their mothers." Embryo transplantation, including cloned embryos, cannot be viable outside of a laboratory or a small, very elite breeding-stock farm. In the real world where the best beef is produced, it will always be done one cow (and one bull) at a time.

.   .   .

Cows and heifers are bred twice a year on the Felton Ranch, which is a little unusual. A lot of ranchers like to get breeding and calving done in one or two months apiece. This concentrates the work, and ranchers like most people like to get a job done and move on. It was traditional in the West (indeed, in the Northern Hemisphere) until recently to breed all the cows in late May or June and have calves arrive the following February and March. It is certainly an ancient practice, for the simple reason that the calf's demand for food, both milk and pasturage, will grow exponentially as the calf matures. In a cold climate, that means timing births to ensure good pasture in the last few months before calves are weaned. As the calves approach the weaning date, around six months of age, they will have good pasture when they begin to graze as much as they suckle. Large wild animals like deer who have shorter gestation periods (and correspondingly faster maturing times) breed in the late fall and have their fawns in early summer for the same reasons—maximum milk and bountiful forage before fall comes and it is time to be weaned and for the doe to breed again.

Richard Felton is growing fond of winter breeding and fall birthing, and each year of late he has shifted more and more toward fall calving season. One of the reasons he started fall breeding was simple logistics: He's the only full-time stockman, and having all the calves come at once was getting to be more than he wanted to handle. Another reason is even easier to understand. There's a huge difference between a Montana spring and a Montana fall—weather. Fall is much warmer; on average, it's a difference of about 30 degrees Fahrenheit higher in the maximum temperatures and as much as 40 degrees warmer overnight (plus 16° and 22° Celsius, respectively). The earth warms up slowly and cools down slowly. A snowstorm on the fall equinox is certainly possible; a blizzard near the spring equinox is positively certain.

When asked why it took him so long to start having fall calves and why he'd stuck with spring calves, he said: "Because Dad did it that way." And the answer to why his dad did it that way was as simple: "He said his dad did it that way." In such a recently settled state as Montana, that is enough generations to move the story all the way back to the first stockmen on the plains.

There are a couple of other advantages to fall calving besides decent weather during deliveries. In spite of the fact that the young calves will be outdoors in some horrid cold weather when they are just a few months old, they thrive on it. Suckling well-fed mothers, drinking out of icy troughs, trying a little alfalfa hay when the mood strikes them, the little black calves are perfectly at home in a frozen universe. Hot weather and the inevitable dusty air and plagues of flies and mosquitoes that come with a Montana summer are harder on the young animals than 30-below weather. This shows up in the catalog for the annual production sale the Feltons always hold the first full week of February. The weaning weight (arbitrarily taken at 205 days, around the time of actual weaning) of fall bull calves averages almost 10 percent more than that of spring calves. At a recent sale the fall calves averaged 680 pounds (308 kilograms) at 205 days, compared to 640 pounds (290 kilograms) for spring bull calves weighed at the same age. Sold at around fourteen or fifteen months of age at the annual February sale, the fall bull calves are full-sized and sexually mature, and when one is sold at fifteen months instead of twenty-one or twenty-two (the typical age of spring bull calves at sale), that's six months of feeding and husbanding saved. In a way, it's a kind of advertisement for the ranch to show animals nearly as large at fourteen months as the average "full-grown" two-year-old spring calves. It just has "growth rate" written all over the animals.

Good as fall calving is for the ranch (and for Richard's ears, the first thing to frost up in March), visitors do hope that he always has spring calves. It's a lot more enjoyable to watch the breeding season in June than in January. But once someone has helped out in both spring and fall, he will understand the reasoning. Chasing calves on foot is part of the game, and it's a lot easier chasing them on autumn grass than on March's ice and snow.

Breeding season any time of the year fascinates. There is no other moment when the eons of domestication seem to roll away and the animals become most natural. Ask the average Montanan what cattle do and he is almost sure to say they eat grass and they poop. (People only semifamiliar with cows, and that's 99 percent of Montanans, always forget the other main activity, drooling.) True enough, but species-ist.

Breeding season, which 99.9 percent of Montanans have never observed, is unique. A bull among the cows regains his dignity. A Mithraist bull-worshiper, like most of the Roman soldiers in the late empire, would say the bull regains its divinity. Richard's mother, cousin Margaret, who was no slouch at judging cattle when she and Raymond ran the ranch by themselves, has a funny way of discussing the Angus bulls on the ranch. When one was roaring away of an evening all by himself in a small pasture, she explained what all the noise was about: "He is expressing his *bull* nature," she said. When he is with the cows, he is *all* bull nature.

The best time to see this is in the pleasant month of June, when meadowlarks perch on fence posts and the sturdier shrubs and sing their lilting territorial songs; when the prairie's few flowers are all in bloom, spikes of soap yucca glowing white, blue wild flax putting out its daily quota of flowers. (They fade after a day and a new one appears. Flax is a renewable resource of beauty.) Even at its greenest, a High Plains pasture is pale, subtle, almost transparent green, a sort of leafy emerald filter that does not obscure the soil or last year's dead vegetation. The black cows are a statement of solidity in an otherwise translucent landscape. And they give a scale to it. Distances are hard to estimate on the plains, but cows (like buffalo or antelope) run to much the same size in a breed. On a hillside that appears to be a short walk away, a seven-iron to a golfer, or a football field and a half, the black cow is shockingly small. It turns out, on approach, to be several football fields to the slope that seemed to loom over the Jeep trail.

Using cows as measurement of distance and height is something that comes naturally, even to tourists from the city. In times past people with large egos have capitalized on this innate human sense of proportionality. James J. Hill had a big ego. He was the magnate of the Great Northern Railway that ran from Minneapolis to Tacoma, right through Margaret's hometown and, more notably, along Glacier National Park. He named a train after his self-image. It's the Empire Builder, and it still runs as an Amtrak legacy. At his already large-enough estate in Minneapolis, Hill was the first American owner of a herd of Irish cattle—Kerrys. The Kerry is black, horned, and the smallest living breed of domestic cattle, averaging well under three feet (under ninety centimeters) in height. Pasturing on his front lawn, they magnified his house by a factor of a hundred percent. English owners of a modestly sized estate favored the Alderney, a diminutive dairy breed that had the same enlarging effect on their landscaping.

The entirely undeceptive Angus cows, each accompanied by a calf, very occasionally by twins, are possibly pregnant (not a little bit pregnant, that doesn't work for cows either). A hundred cows that have had one shot of artificial insemination, grazing in a large pasture, say, a square mile (260 hectares), are enough to keep a bull occupied sorting out which thirty-odd open cows need his services and which don't. His cows will be scattered from fence to fence, occasionally coming to water, but not flocked or herded up. Angus will scatter over the landscape, picking and choosing among the grasses and forbs. A good Aberdeen-Angus mother will cover miles in a day if necessary, keeping her milk supply prolific.

Some animals that share the ranch with the black cows create tightly bunched harems in breeding season. This is a reasonable strategy for mule deer and whitetails and antelope. Those female animals come into season just once a year in the late fall during the male rut; collecting and guarding a harem is one way for the dominant buck to impose his genetic will on the herd. Smaller bucks will skulk around the fringes, and sometimes succeed in breeding furtively while the big buck's attention is distracted.

This doesn't work for cattle. With females that come into heat once a month until they are pregnant, and with bulls in a state of constant passion, there is no need for and no utility in a tightly guarded harem. In a state of nature, females would cycle shortly after giving birth (if food was easily available and nutritious) or some months later, when they had recovered from the drain of pregnancy and nursing and food was seasonally more abundant. Their collective breeding season could be scattered throughout the year, although in rugged climates, breeding would tend to be in the flush grass season of summer, with calving season about two hundred and eighty days later on the cusp of spring. Rather than creating a harem, the aurochs and his descendants would guard a much larger territory and make daily diagnostic visits to the scattered cows.

The well-fed cows on the Felton Ranch return to their monthly cycling almost immediately after calving, and they'll be bred again within two or three months. Whether in the pasture or in the primeval forest, the bull's work is to find the female in heat whenever and wherever she is and breed her.

On a ranch, whether artificial insemination is used or not, the bulls are kept distant and separate from the cows for most of the year, loitering in their own bull pastures. Absent nearby cows in heat, and with no one to play with but one another, bulls spend most of the year quietly. If there is something interesting to butt around (a hay bale or a mailbox), they will destroy whatever it is and spend the bulk of their daylight hours eating, sleeping, and confirming the bossing order with an occasional head-to-head pushing match. Then, nine and ten months before the rancher wants a crop of calves, the bulls are brought to the heifers and cows.

As we noted, while it is fairly easy to identify the AI-ed calves and the naturally bred calves by their birth dates, there can be problems because an extensive ranch is not a perfect place. Laboratory conditions are unavailable. Not only do fences go down, mixing a few cows in with their next-door neighbors, or letting another bull into the pasture, but a full-grown Angus bull, if it has a mind to do so, will jump a three-wire barbed-wire fence. This looks impossible, given the animal's bulk and its usual slow-walking demeanor. It even looks impossible when it happens right in front of you. It is one of the mysteries of the ranch world that bouncy, springy antelope will not try to jump a fence, although they can bound that high and higher fleeing from some threat. Antelope sneak through fences like four-footed limbo dancers. Range bulls, who look like they couldn't jump over a bucket, will soar for sex.

Even after invading a pasture, an uninvited bull has little chance of breeding. He is by definition starting at the bottom of the bossing order, and the resident bull has all the psychological advantage. To be sure, there are bulls so dominant that all pastures are the same to them, but these are uncommon beasts. To see an interloping bull loitering helplessly and hopelessly near the dominant bull is to understand exactly what the common English word implies. He looks, he acts, he is "bullied." He may not even seem interested in breeding. This is not necessarily a ruse; it may be a result of losing a couple of shoving matches with the resident bull. In mammals, including humans, defeat in a physical contest produces an immediate and steep decline in circulating testosterone. The elegant research has been done on collegiate wrestlers, but it appears to be the same with bulls. To lose is to lose interest.

Once the ranch bull is in with the cows, if he is a good bull, he will do what all bulls have done since aurochs ruled the forest. Several times each day, he will visit every cow and assess her state. The pheromones of heat and approaching heat come to his large nose through the air. He will spend his day walking from cow to cow (at a purposeful pace that looks slow but would wear down a human pedestrian). He will smell urine, he will smell butt, and this is all done in a businesslike manner. Tourists to Montana may see an isolated bull or two in the summer months. By himself he is stolid and slow, moving only enough to come to new grass or to stroll over to the water trough. With work to do and almost never with nonranch spectators, he becomes another animal entirely—active, purposeful, dominating.

A bull testing for pheromones

As one drives through the pasture in breeding season, it is possible to think that a very large cow lying down or otherwise concealing her udder and her feminine head might be the bull. But there's no mistaking a bull for a cow. When he comes up out of a draw and through the sagebrush, there is a set to his shoulders and a deliberateness, an invincibility to his progress, and it becomes immediately apparent that this is the bull. Bulls do not strut or swagger. Bulls are mass in motion; they have momentum.

Breeding season is also the only time when bulls have something like a real expression on their face. Like many other large quadrupeds, the searching bull breathes with his head held high and his upper lips curled open as he inhales so that the air passes over his exposed gums. It produces the impression of something between a sneer and the flared-nostrils and drawn-back upper lips of someone about to go berserk.

What the bull is up to, however, is not a true expression; rather, he is adding another sensory organ to the search for the breedable female. Toward the back of his palate, there are a pair of organs, one on either side, that open into a void lined with an olfactory mucosa that is especially sensitive to the sexual odors of the cows in heat. It is well supplied with nerve endings. In effect, the bull has a second smeller, as truly sensitive as the mythical human second sight. This act is called *flehmen,* from the German word for exposing the lip and looking like Hermann Göring. The other thing that a dominating *flehmening* bull will do is flex his muscles. Bulls do exactly what human bodybuilders do on their stage: The bull can make all of his large shoulder muscles bulge at the same time.

To an anthropomorphizing eye, there are occasional moments that look like courtship. If the cow rejects the bull or if he determines she is not yet ovulating, he must be away and about his duty, examining the other far-flung cows. But once a bull has started to follow a particular cow, sensing that she is about to release her egg into her uterus, where the sperm can contact it, he will be quite attentive. If she lies down, he may join her, lying close enough to touch her, back to back or head to head. (Cows, as noted before, are contact animals by nature.) If she is lying calmly, he will drop his enormous head gently upon her body. (Swain's head in the damsel's lap would be the human equivalent, but quadrupeds do not have laps.) With just a few cows coming into heat each day and with temperate weather, the breeding season occasionally has this bucolic sensibility. It seems there must be all day to loll about, so placid are the animals. But these moments are short-lived, measured in minutes, not hours. Most of the day it is loitering cows and a perambulating bull.

A bull in flehmen *posture*

When she is ready and he mounts, his penis becomes erect and he rather frantically probes. When he enters, the rear of his organ, lying curved inside his abdomen, straightens. It has been folded in a loose *S* shape, held under considerable tension by the retractor penis muscle. When the muscle relaxes, the penis straightens and thrusts and immediately he ejaculates. His penis spasms at the time of ejaculation and the business end oscillates, spraying sperm all around the cow's rather capacious vagina. The sperm have about twenty-four hours to pass through the cervix and another full day to natate in the womb, enter and search both of the pair of uterine horns, and find the descending egg.

Except for the laying on of heads, there is no recognizable foreplay. Even though the cow has all the characteristic mammalian external sexual characteristics necessary for excitement and orgasm, they seem to be superfluous. She is not interested in titillation or repetition. This is all to the good from the evolutionary or the modern economic point of view. There are more cows to be checked and miles to go before the bull sleeps. The last thing a rancher or Nature wants is a nymphomaniac cow monopolizing a bull's time or a bull that doesn't know when to quit. Richard had a bull once that just wouldn't stop copulating or trying to, whether the cows were open or not, and although the bull had otherwise excellent breeding and did turn out good calves, he went right along to the packing plant in Miles City.

→10←

"Mothering Ability" Demystified ... Almost

SOMEONE driving by a herd of pregnant cows some nine months after they were bred, even if one or two of them were about to go into labor, would not see anything remarkable. But people working with cows and looking after them in all seasons would see signs of impending births. Cows can scratch their bellies with a hind hoof, so seeing them standing three-legged and poking at their tummies is no surprise. But in the first stages of labor the cows aren't scratching; they're actually kicking themselves—gently but still more aggressively than a mere scratch. They are restless, and that is unusual in the animal. A restless cow is not a happy or healthy cow. They start to walk a little stiff-legged and a little splay-footed, their hooves falling outside the drip-line of their thick bodies. They will even take a step or two backward, which a cow almost never does by choice. Even when crowded like a lei of black orchids around one of the ring feeders, once they're through feeding they will try (and usually succeed in) getting out of the crowd by swinging their heads to one side and walking almost straight but a little crabwise, rather than backing out of the feeding mob. Cows will back up from a butting superior or if someone is trying to herd them from one corral to another. They will back up in a chute when the rancher is trying to get them forward into a workable position. The cow being ordinarily such a straight-ahead animal, seeing one moving backward on her own attracts attention.

Inside the cow, powerful chemicals are starting to circulate in greater volume, all of them basically identical in man or mouse, cow or rhinoceros. They are common always-present hormones. But not only is their quantity increasing; more important, cellular receptors for them are multiplying exponentially in the cow's brain, her uterus, and her milk-producing glands. The most powerful of these hormones for everything that is about to happen is oxytocin, the very first peptide hormone ever identified and synthesized, known to science since the turn of the twentieth century. On a cattle ranch in calving season, there may be vials of oxytocin in the pickup truck or the saddlebag in case the rancher wants to accentuate one of the natural processes about to occur by dosing the cow. It is used much as it would be used in a hospital obstetrical unit, to begin contractions or to accelerate contractions if the placenta is not being expelled. The needle is not usually necessary. Several tens of millions of years of mammalian evolution have provided the average range cow with all the oxytocin she needs. In fact, even as the birth contractions begin, her oxytocin levels rise very slightly; less than 10 percent more of this peptide circulates than has been in her blood during the nine months of her pregnancy. It is all those new receptors or pathways on the cell surfaces that are the great change.

The new receptors suck up the oxytocin with very different results, depending on the cell. In the uterus the oxytocin stimulates the first contractions to expel the calf; in the mammary glands, it both stimulates milk production and allows the first flow of milk to the udder (what stockmen call letting down the milk). The largest proportional increase in oxytocin receptors is in the uterus by a factor of at least two hundred new receptors for every one previously present. In the brain oxytocin receptors suddenly flourish in very specific areas. One sector of the brain with a very large increase in receptors is the area devoted to the sense of smell. In her brain, besides the nasal center, receptors sprout on neural cells in the area that is responsible for maternal behavior, for attending to the calf. Other receptors develop in areas that control her fear of new things—areas that, when stimulated by the hormone, allow her to explore new surroundings, to inspect new objects without anxiety. And oxytocin serves as a facilitator for the production of dopamine, a hormonelike substance associated with mothering behavior and finding pleasure in the activity.

The scientific investigation of these matters tends to take place, like everything else, in laboratory mice, but the results seem to apply very exactly to other mammals, including us. Virgin mice receiving a shot of oxytocin will start collecting nesting material. Given the opportunity, they will start kidnapping another mouse's pups. Mice of both sexes, given elevated levels of oxytocin, show much more willingness to explore unfamiliar areas of the laboratory floor, mazes, hiding places, even new toys. Stockmen tend to talk of mothering ability, and call it an instinct, as if the cow was hard-wired to nurse and protect the calf. Well, in a way the cow is predisposed, but it requires hormones like oxytocin to provide the metaphorical electricity to enable the hard-wiring to do its job. This is an incredibly efficient, elegant, piece of mammalian engineering. A single peptide triggers contractions, begins milk flow, predisposes the cow to accept a calf (particularly important with heifers, who have no idea of what's just happened to them), enhances her ability to care for that calf, and gives her pleasure for doing so.

The role of oxytocin in facilitating exploration and acceptance of new objects can be critical to calving success. Heifers, having their first calf, may act as if they haven't the foggiest notion what this slimy black critter is, how it got there, and what to do next. As cows usually give birth lying down on one side, they instinctively look back over the upper shoulder toward the calf, even before rising. When watching a heifer approach her calf after she is up on her feet, there are clear indications in the heifer's movements that suggest she is doing so at least as much out of curiosity as out of caring, maternal interest. She may even step backward in what looks like a startled movement if the calf moves. But the heifer is willing to try to deal with this thing on the ground. As the mice high on oxytocin will accept a new toy without anxiety, so will she. (One is tempted to say "have affection for it." When someone is watching it happen, seeing the intensity of the bonding, "affection" seems a reasonable word.)

It is difficult to tell if labor is painful for the cow. There is some odd-sounding mooing and some deep breathing, but the cow is a stoical animal in all seasons. And when the oxytocin, in addition to all its other duties, triggers the release of dopamines, it is giving the cow something of a light, general anesthetic, a palliative against pain.

The first sight a rancher gets of the calf will come while the cow is still standing, when a pinkish-gray ball of amniotic-sac-covered calf is extruded a few inches out into the world. If things are going well, the first part of the calf's anatomy to show itself will be the front hooves. Small and sharp-edged, they push visibly against the sac and are the best of all possible signs. Hooves that face forward with the bottoms toward the ground show that the calf is oriented to come out headfirst and right-side-up, its hooves leading the way, like someone diving head-first, hands out and palms together, to facilitate a smooth entry into the water. While the cow is still standing, you can imagine the calf diving for the ground. But she will lie down and in a few heaves of her side, especially with an older and experienced cow, the calf comes fairly popping out. When the cow's water breaks, emptying the amniotic sac, it is as voluminous as you might expect from a cavity holding a calf that will be nearly three feet long and weigh upward of a hundred pounds (80 to 90 centimeters in length and weighing as much as 45 kilo-grams). One moment there is a pair of legs and maybe the first outline of the skull, and the next—if you blink you'll miss it—there is a whole calf lying on the ground. Not only does the calf come out more easily when it is head and hooves first, but the amniotic sac is usually punctured by the tiny hooves and tends to peel back from the head. This uncovers the nose and mouth of the calf so that it can immediately take a breath.

With heifers, that is, females who have not calved before, the combination of hard-wired instinct and the sudden rush of hormones can result in too-eventful births. Dramatic mo-ments are not encouraged in the cattle business, any more, for example, than serious explorers want to have adventures. The way cattle are raised, separated by age and sex in different pas-tures, means that heifers won't have any role models or help-ing sisters when their time comes. This is probably just as well, because older cows can be too interested in what's happening, and some will try to "steal" a heifer's calf.

As an example of what can go wrong, on a recent fall day at the Felton Ranch, a clear and comfortably cool day, a heifer's labor started out badly. Her calf was coming out backward, and the hind hooves, pushing against the amniotic sac, were sticking out about a foot, but the rest of the calf wasn't following. A calf's butt is not only larger around than its head, but it is squarer, bad geometry to ease its way down the birth canal. This is hard enough for an experienced cow, with a stretched and pliable uterus. It's a chance for tragedy in a heifer. So it was that she had gathered an audience, just in case things got worse and the calf needed to be extracted. But this calf came without human help, butt first and all at once. The heifer got up, backed up a few feet, and stared at the bundle of slimy hair and slippery sac as if it were some kind of extraterrestrial creature. And then, you could just almost see the oxytocin kick in as the heifer moved quickly up to the calf, and as all good mothering Angus do, she started to lick the calf vigorously.

This is when it helps to have a human being on the premises. There is no part of a calf, from birth to weaning, that interests the mother more than the calf's anal region. And this calf, with its hindquarters sticking out of the broken sac, began to get a thorough licking-over on its butt. The sac, however, was still solidly in place over its nose. Richard took one look at what was going on, muttered something unprintable, and hopped the three-wire fence, except that his right boot caught on the top wire and redirected his considerable momentum from horizontally forward to vertically downward, creating a situation in which all the unprintables were perfectly audible. He pulled the sac off the calf's head and stepped back and watched. This was a good day, and the calf coughed up the fluid it had inhaled and started to breathe shallowly but regularly. (It can't help breathing in some fluids. It can't hold its breath until help arrives, whether it's a rancher's hands or a mother's tongue.) On not good days, picking the calf up by the heels and shaking it may dislodge the aspirated fluids. That's not as easy as it sounds, because the calf is slippery and heavy. This is one of many moments on the ranch when the rancher hopes his Angus cow has good mothering instincts but not too good. A little excess protectiveness and the midwife will be attacked.

From their personal experience and anecdotal evidence, Aberdeen-Angus stockmen believe their black cows are superior mothers. Scientifically, there is little research, but what there is indicates that their hands-on wisdom is real, not just wishful thinking. In one experiment, observers watched cows for the eight hours immediately after they calved, comparing dairy cows to Angus mommas. Dairy heifers were barely attentive, and when they did get around to licking a calf, they accumulated, on average, just eleven minutes' licking time in that eight-hour day. Mature dairy cows spent thirty-two minutes licking in the eight hours. Angus were notably more interested. Both cows and heifers started licking immediately and averaged forty-eight minutes of tongue time in eight hours. That the heifers were so attentive is truly remarkable. In another study, confined to beef breeds, researchers separated the calves from the cows and rated the cows on their attentiveness on a scale from 1 (aggressively attentive) all the way down to 11 (oblivious). Angus averaged in the low 5s (the upper end of attentive), all the other breeds in the 6s (somewhat attentive). The studies were done in Europe, where considerably more handling and other human contact occurs. Having someone borrow a calf might be tolerated more there than in a large-scale ranching situation. A herd of range-wise Aberdeen-Angus would average even closer to "aggressive"; certainly they would score "highly attentive."

As mentioned earlier, the increase in oxytocin receptors in the olfactory center enables an increase in the mother's interest in, and ability to remember, the very precise and unique smell of her calf and, specifically, her calf's anal region. The frequency of this smelling borders on the obsessive, and whenever the calf has strayed and then returned, or the calf tries to nurse, or a strange calf tries to nurse, the mother's nose is unerring in identifying its babe. This makes it difficult to transfer calves from one cow to another, something that has to be done when the mother dies soon after giving birth. More rarely, particularly with good Angus stock, a mother will actually reject a calf, a behavior that will end her career.

To make a pair-bond of a strange calf and a strange cow (or an orphaned lamb and a ewe that has lost its lamb) requires use of one of the oldest tricks in animal husbandry. To replace a dead calf with a living one, and it works equally with sheep, ranchers will skin the dead one and tie the hide onto the orphan. Even after the calf's death, the cow can recognize that unique odor (probably a blend of many odors) that tells her this calf is hers. She can continue to pick out the behavior-triggering odors even as the tied-on hide begins to turn bad and have, to a human nose, no odor but decay. This still is done on the ranch, although Richard has figured out that you don't need to skin the whole dead calf. Just skin out the hindquarters, being careful to include the anal (and vaginal, in the case of heifers) area.

Things rarely work out so that there are equal numbers of bereaved cows and orphaned calves. Some Aberdeen-Angus cows are exceptional milkers, but almost all of them would be able to feed a set of twins. About the only way to get a cow to accept a second calf is to confine them together, with the mother haltered and immobilized. Usually, if she can't butt the orphan away, she'll give up and accept it.

As the hormonal rush of birth subsides, the cow still receives a regular oxytocin high. In all mammals, the act of nursing, facilitated by oxytocin levels that stimulate the milk glands, will in

A cow nursing a calf

itself increase the production of oxytocin. It is a feedback loop
of hormonal generation and application. And oxytocin facili-
tates the production and absorption of the opioids that enhance
the sense of well-being in the mother. This is more obvious in
humans than in cows, because cows have virtually no facial ex-
pressions. That blissful look on the face of the nursing human
mother is of course in part engendered by the mental satisfac-
tion in providing for the child, but it is also generated, and pow-
erfully so, by pulsating levels of oxytocin massaging the pleasure
centers of the brain. Human subjects observed and tested while
their oxytocin was monitored closely responded to the pulsing
increases with very specific behaviors that demonstrated a desire to please, to give, to interact socially. Apes, with a more limited repertory of behaviors, still showed increased levels of overt
mothering, including cleaning and cuddling. And they made
fewer ugly faces directed at the human observers—this is called
a "reduction in agonistic responses" by scientists.

Nursing Aberdeen-Angus cows do seem calmer, even relaxed, when they have young calves around them. However, there is no reduction in agonistic responses if you try to get between them and their calves. Perhaps it is just by contrast with the young ones, who are constantly trotting around looking for another calf to play with, that the mothers lying in the shade at the edge of the pasture look so contented. But there is much more to being a good cow than just the mental state triggered by hormones. There is something about mothering ability that seems to go beyond a mere bundle of bovine instincts, that appears to have elements of socialization, to be influenced by learning.

Cows having a second, or later, calf are likelier to take the calf off into some concealment and may even be more aware of how to overcome a problem. One fall calving season, while Margaret Felton was entertaining a visitor, she noticed a cow in the pasture behind her house that was having a hind-feet-first delivery. Ranch wives see these things, at least the good ones do. Richard was a couple of miles away tending to some calving heifers, and it looked as if there was no particular hurry. The labor had just begun and the water wasn't broken. Margaret got the cow's number from the ear tag just in case, as does happen, the fetus slipped back inside and one couldn't see the hind hooves. (Richard, like most breeders, wouldn't need the number. He'd recognize and remember the cow without even thinking about it.) By the time he got back to the pasture with the problem birth, there was a healthy calf on the ground. It is a good cow that can pull off an unattended backward birth and get the calf breathing. Richard, who doesn't get emotional about any of his cattle, remarked as much: "Takes a good momma to pull off that trick."

Much as farmers and ranchers think well of them, there are

some ways in which Aberdeen-Angus cows, excellent mothers taken as a whole, do seem a little oblivious of reality. Like most animals, they will hide their recently born calves and leave them while they go to water, graze, or socialize with the other cows. Their talent at hiding is somewhat less than their other mothering abilities. Fall calves, coming in September, are often born on the alfalfa fields around the ranch houses, and that's where the mother hides them. But the fields have all been cut two and three times, the alfalfa is not more than six inches high, and the calf, lying flat, is just a little less obvious than a mouse on a pool table.

In the spring calving season (which is really the late-winter season), the herd will be in wooded bottomlands where there are much better places to hide a calf, and the cows will use them. When looking for newborns in the morning, one looks in the brush along the ditches and creek bottoms, and the first thing seen, more often than not, is a very sober and serious Angus cow looking directly at the intruder. After that, it is easy to find the calf.

When a new mother goes to socialize, it is all the usual stuff. She will get involved in the day's butting (or bossing) ritual, maintaining her place in the order of things, something like a human mother going back to work part-time or having the neighborhood wives over for coffee. The greatest amount of head butting or boss shoving takes place at the ring-feeder (circular metal containers into which hay is placed to keep it dry and clean until it is consumed), where one of the half-ton round bales of hay is placed to supplement the alfalfa stubble and the wild grass outside the irrigated fields. When cows do leave their newborns, they typically hide the calf (and if they are in a wooded, brushy pasture, hide them very well). As time goes by, cows will leave their calves with an "auntie." In wild-animal behavior this is termed making a crêche. In Britain, the word "crêche," adopted from French, is commonly used to describe a day-care nursery. American English uses it only as the term for a statuary representation of the Christian nativity. Men who have been associated with Aberdeen-Angus cattle in Scotland over the years are familiar with crêche behavior, but two of them remarked that it was increasingly unusual, if not entirely absent, in the strains of Aberdeen-Angus cows being raised there today.

On the ranch, mothering ability means more than bonding with the calf and suckling it. That might be all you need in a small farm or a typical British breeding stock operation. With range cattle, it becomes a complicated set of behaviors, and like so much else about the business of breeding cattle, it is a question of balance. Take milk production: A cow with too little milk or too little tolerance for nursing will raise a smaller calf than desired. But a cow with too much milk can raise a calf too fat for its own good. Overfeeding calves (including feeding too much grain to weaned calves) will get you a bull calf with small testicles and a heifer with delayed puberty. This is not a problem when cattle are kept indoors as they are at dairies and most British breeding-stock farms. There, the young calves are penned separately from the cows and allowed to nurse just twice a day for an hour or less. With range cattle, the stockman has to depend on the mother to regulate the amount of nursing and on the calf's instinctive, accurate sense of satiety.

A story from the Feltons' past illustrates the problem of too much milk. Margaret, in a conversation about the steers being fed up for family use, remarked on a particularly good beef animal they had raised and eaten. Better than good, she said, it was the best ever. Surprisingly, the animal in question was not an Aberdeen-Angus. It was a milk cow, a Jersey heifer that got too fat to ever milk decently. What had happened was that they had an Angus cow that had lost a calf and a Jersey cow that went down shortly after giving birth. They decided to put the little Jersey calf out with this bereaved Angus. (The alternative was tedious hand-feeding or bringing the Jersey calf to its mother for a few hours a day.) When that calf got a chance to suckle every daylight hour, she blew up into an excellent beef animal, useless for milking.

The European dairy breeds are unique in the cattle world for one trait: They will continue to produce milk without the stimulus of a calf at their heels. (The earliest representations of human milking cows, from the Middle East or Egypt, always show a calf at the cow's side, with the milker working from the stern.) The ideal dairy cow is completely oblivious of her calf; after all, the bull calves and most of the heifers will end up as veal. Like differences between types of cattle, mothering varies among the dairy breeds. The older style of breeds, ones less genetically manipulated for nothing but voluminous milk production, can be very good, very attentive mothers. When observers note the poorer mothering skills (and interest) of dairy cows, it is usually a herd of Holsteins they watch. Jerseys, the smallest of the commercial dairy cattle (and the least common today) are typically excellent mothers, which along with the high quality of their milk can make up for the shortfall in pounds of milk per cow.

Calving season is when ranchers really get to know their animals. Angus, by and large, are seldom mean. They certainly aren't wild. But occasionally, and attention must be paid when it happens, there will be a "ringy" one. A ringy cow is one that takes a natural instinct and pushes it too far. A cow that takes you when you try to grab her calf is ringy. So is a cow that wants too much to be alone when she calves and decides to come out of the sheltered bottom where the rest of the herd is lying low in a storm and have her calf out in the open prairie with the temperature well below zero (–18° Celsius) and the wind up near a gale. Nature would punish this behavior and leave her staring at a dead and frozen calf. On the ranch, that's not an option, and ranchers end up playing midwife in a blizzard with blood and amniotic fluid freezing on their hands and stiffening their clothing as they try to quickly dry off the calf. There are special-made calf blankets, like little wool-felt sleeping bags for calves that can hold in some body heat while the calf is taken in out of the weather. When it's breathing easily and warmed up to a normal temperature, it goes back to Mama. Most of the time even the dumbest cow will accept her calf when it's brought back.

Even in these almost civilized days on the High Plains, a mothering instinct to defend the calf still matters. The wolves haven't made it back to most of cattle country, but there are always coyotes. A few springs ago they got into a calving pasture on the ranch. Even though the cows and new calves were less than a quarter-mile from a house, less than a hundred yards from a ranch road, coyotes attacked, and the morning revealed one half-eaten calf and one with four distinct puncture wounds high on its right hind leg, just matching a coyote's upper and lower fangs. That calf survived. The hide around the punctures did slough off, but it regrew without a hint of what had happened. It takes little imagination to re-create the night of the coyotes. One mother stood and watched; one mother got between the coyote and the calf. Even without horns Angus cows can be dangerous. For one thing, they have surprisingly quick feet and could snap a coyote's back like a No. 2 pencil.

Such protectiveness is all to the good, except for the obvious problem. Calves must be handled almost as soon as they're born. On the ranch Richard needs to record their birth weight so as to calculate their growth rate (and also to make sure he is not breeding up animals that are dangerously large at birth). He has to punch a tag through their ears with a unique number that will always identify them and keep their genealogy perfectly clear. And the rancher has to do this with a half-ton cow staring at him, maybe even blowing snot and scraping the ground with one of those too-quick front feet. And most of the time, he'll have to grab the calf as it flees away from him and the cow, and he'll end up working the calf with the mother behind him out of sight. It doesn't happen too often, but the occasional cow will insist on getting the calf back, and right now. One advantage the stockman has is speed, not foot speed but quick and efficient ear-tagging and bone-measuring. Cows do take their time making up their minds what to do next, and after the work is done, the cow's usual reaction is to take the calf and clear out. And the other advantage the human has is timing. If the cow is really rank, the rancher can come back later and see if it hasn't calmed down, which it usually will have done. Calf-tagging is one of those times that fosters speculation on the difference between a domesticated animal and a pet animal.

The Angus equivalent of emotional maternal concern, call it what you will, is evident when a calf is stillborn. From a hundred yards distance, the calf is visibly dead. It is just a little bit too flat against the prairie; it is flaccid, even more so than a sleeping calf. It looks like a discarded toy with the air let out. The mother will lick it clean, smell it repeatedly, even prod it with a hoof, or use her nose to roll it up onto its belly. And that may seem to be just the hard-wired instinct to perform some simple mechanical tasks. She will wander away from the diminutive corpse. She will come back, and that is all wiring too. And then she will begin to wail. Her posture is the same as that bulls adopt when bellowing. She holds her head up parallel with her back, not drooped at the usual comfortable angle. She arches her back, stiffens her hindquarters, and howls. To what and exactly why are mysteries. But it is hard to watch. It looks like a great sense of loss, and it sounds like any mother keening. It will be a few days before she stops coming back, returning to see if the calf is . . . what? Ready to nurse? Ready to be licked? Ready to be led away to a quiet place in the nearby brushy bottom? Maybe none of those things, maybe it is just an instinct that needs some time to extinguish itself. But there is still the wailing. What could possibly be the genetic cause of that?

The Importance of Herd
Behavior to Cattle

IT was noted earlier that Aberdeen-Angus cattle have something bordering on personality, that they are on the edge of, when appropriate, a sense of self-worth. An oddity in the way they sort out dominance in a herd is yet another example of the uniqueness of the breed, another indication that they have a palpable self-esteem. All cattle will build a social order within the herd. It has the same effect with cows that it does with, for example, free-ranging chickens.

Dominance, for lack of a clearer word, is a way of reducing antagonistic behaviors. In the pecking order of old-fashioned barnyard chickens, it could be described as a scale from one to infinity, where the number one chicken could peck any other chicken with impunity and the last chicken couldn't peck anyone. As little family flocks of chicks reach maturity, they are gathered into the larger flock's pecking order and find their level. If chickens didn't recognize one another, and couldn't remember who got to peck whom, it would be financially unfortunate for the farmer and horrible for the last chicken in the order. But once dominance is acquired, antagonistic pecking almost disappears and the occasional bout between two chickens is brief. As a corollary, when chickens are raised unnaturally, kept together in groups of the same age, and crowded in rearing barns, there is neither time nor space (for the loser to disappear into) to develop a pecking order, and fowl are routinely debeaked, the upper mandible being snipped off. These factory-raised birds are not a flock; they are a seething inchoate mass.

In cattle, the acquisition of dominance by individuals produces a social order in which no excess energy is expended arguing over who gets the first shot at food and water. Dominant cows and bulls get first pickings. Even with fifty cows trying to get at one or two ring-feeders, there is surprisingly little pushing and shoving. Dominance in cattle is the equivalent of table manners in human beings. Unlike chickens, cattle start sorting out the order when they are very young. Aberdeen-Angus calves just a few weeks old are constantly playing with one another. They have what appear to be footraces and a kind of wrestling that imitates adult sexual behavior by alternatively mounting each other until one runs off. Very early in life, butting or shoving each other around is the most common interaction or, from the looks of it, their favorite game. They don't run at one another; they carefully move in close, make head-to-head contact, and see who can push whom backward. In a few weeks or a month, the amount of butting among the calves is noticeably reduced and, by the time they are grazing as well as suckling, extremely occasional.

With some domesticated animals, behavioral scientists may argue that the human being simply takes over the role of alpha male or alpha female and is similarly dominant over, for example, a pet dog. It is not at all like that with cattle; the cowboy or the herdsman is not at the top of the pecking order. What control he has comes from taking advantage of the dominance hierarchy whenever possible and working around it when necessary. Take an obvious example—moving a herd from one pasture to another. There is no point in trying to get the herd moving by trying to impose some leadership on them, picking out a cow or a bull calf or whatever to lead the way. One of two things is going to happen, and there's no use trying to control or speed up events. If the dominant cow wants to go (or doesn't mind going), she will lead the way. This is something dairymen see twice a day when the cows come into the barn on their own to be milked. The dominant cow is always first in line and the least dominant is at the very end. Contrarily, if the dominant members of the herd, the top cattle, don't want to go, then the least powerful animals will finally respond to the urging and lead the way. And since they are herd animals, the dominant cows will reluctantly trail along, bringing up the rear. The same is true with a group of bull calves or heifers: The choice of leader of the pack has been made by the pack and there's no changing it.

While sorted-out dominance makes for a happy herd, it can cause problems for the cattleman, some serious, some just annoying. Weaned bull calves are the most bumptious herds on a ranch, part fulminating hormones, part youthful insouciance. This makes it difficult, usually unwise, to try to integrate two established herds into one paddock or introduce a single outsider. A bull calf on the Tongue River ranch came up with either a broken shoulder bone or a badly dislocated bone and had to be taken out of the bull calf pasture before he got hurt even worse. He was turned out on his own in the pastures, including twenty acres of irrigated alfalfa adjacent to the bull calf paddock, and eventually healed himself, either remodeling the broken bone or getting the dislocated bone back in its proper socket. But Richard didn't dare put him back in with his former herd-mates. He'd been gone too long, had no place in the pecking, or butting, order and would, Richard said conclusively, really get the crap beaten out of him. Being sociable to a fault, the solitary bull calf kept trying to get back in the pasture with his classmates. Once or twice a day someone would see him working away the fence and have to chase him off. He only succeeded—when he was fully healed—in attempting to jump the four-wire fence, catching up on the top wire and effectively ending his career as an impregnating bull.

At the annual Felton production sale, this herd clannishness gets expressed in the sales catalog, where animals are listed and where the buyers expect the order in which the stock comes into the auction to be exactly the same as the printed order. Bull calves come in two age groups: bull calves around fourteen months of age (born in the fall a little over a year before the February sale); and two-year-old bull calves (born in spring, twenty-one to twenty-three months before the sale). And it makes common sense to group them separately. But bull calves also come in another distinction: Some are raised on the Tongue River and a smaller number at the home ranch

Getting a bull to go through a chute

on the Yellowstone. And they have to be kept apart. So what looks like a simple progression through the catalog is actually a progression in four groups, matching the four groups that were raised apart from one another. As they are held in separate pens outside the auction barn, to prevent excessive shows of aggression, so they can be brought into the barn only one at a time from the same pen or same herd.

A constant annoyance caused by the Aberdeen-Angus's ability to form a strong, clear dominance order comes when it's time to work a herd of mature animals through a chute for vaccinating or some other routine task. Imagine a single herd of pregnant cows, numbering nearly three hundred, and all going into a chute one at a time and then down the chute into the squeeze cage. It is the same principle as trying to move the herd as a whole—if they don't want to go, they go in the reverse order of their dominance, wussy cows at the beginning, boss cows last of all. This wouldn't matter except that running three hundred cows through a chute and working them in the cage takes a long, long day, and toward the end, when everyone is tired and sunburned and thirsty and maybe even grumpy, the last dozen or so cows are dominant, bossy, tough cows, and they *really* don't want to go into the squeeze chute. It can take as long to work the last ten cows as the first fifty, and it feels like it's taking as long as the first one hundred. It's almost funny, knowing that every one of the last cows is going to fight it, and every one of them will require considerable physical contact, including some serious tail-twisting, to get them into the squeeze chute. But it is not funny.

The oddest thing about the uncooperative cows, the last ones that will go into the chute, is that they don't look a bit different from the easy cows—they're not any bigger or wilder or spookier. It is not possible to look at a herd of Aberdeen-Angus cows and guess which ones are the top cows. With some other breeds, you can pretty much tell. Dairy cows, which are the most closely observed cattle both by dairymen and animal-behavior scientists (ethologists), sort out their dominance almost entirely by simple mass. The dominant cow will be taller and heavier, and so it will sort out down the line to the smallest cow. In a mixed group where some have kept their full horns, that will add to their dominance quotient, and a medium-sized horned milk cow may dominate a larger dehorned herd-mate. It is hardly so simple with beef animals.

Observers watching a mixed herd of Aberdeen-Angus, Hereford, and Shorthorn cows discovered that all of the (hornless) Aberdeen-Angus were dominant over any of the horned cows, regardless of size. And within the elite circle of black cows, size didn't seem to matter in the social order. Small cows could and did dominate larger and heavier animals. Something invisible, like attitude or some other psychological attribute, seemed to make the difference. But dominance aside (and it is functionally put aside by the cows, once it is settled), the same observers noted that the Aberdeen-Angus were more gregarious among themselves than the other cattle, always more closely spaced than the other beef breeds.

While a herd's natural, healthy, bossing order can cause extra work for the rancher, he cannot and usually does not take it personally. The difficulty has nothing to do with him. The cow that won't go in the chute without effort is just being herself. There are, however, some animals that are mean or sneaky or just plain wild, and in those cases the rancher has to take it personally, if for no other reason than his own safety.

In some ways the most dangerous animal is one that seems to be spontaneously and suddenly aggressive. One of the oldest recorded examples of a sudden outbreak of nastiness comes from, of all places, Hugh Watson's herd at Keillor. Watson had always stressed amiable behavior in his herd. His brood cows were harnessed to plows and seed-drills in planting season, so manageable were they. Even at Keillor the occasional young animal could act as badly as any unruly teenager.

Watson's daughter liked to tell the story of a Monsieur Dutrône, the Secretary General of the French Humane Society who made a pilgrimage to Watson's home to bestow a medal on the farm's longtime herdsman (whom she does not name—even on cow farms there's a class system in Britain): "'for 40 years of tending these harmless and hornless cattle. . . .' Immediately after the medal was bestowed," Miss Watson wrote, "M. Dutrône had to exert his agility to the utmost to escape an onslaught made upon him by a dodded heifer into whose loosebox he had rashly intruded."

"Loosebox" is a term peculiar to Britain, where cattle were not only kept indoors but kept tied to the front of their stalls for the winter, with just enough slack to get up and down, that is, *not* loose. A loosebox was what we might simply call a pen. What Miss Watson termed an onslaught, when translated into American ranch-speak, becomes "trying to take you." The rancher's idiom implies a little sneakiness, as being taken in a business deal or by a convincing sob story. It also suggests the person taken was not paying attention.

There is a significant difference between a corral full of sociable (with one another) heifers and a solitary animal, like the one that went after M. Dutrône. Cattle in a group feel much safer than a single cow. One never hears of a ranch hand or a visitor being charged by a herd of Aberdeen-Angus. If there is a problem, it is with an isolated beast. Given a choice, cattle would rather interact with one another than confront or cotton up to a human. However, it can happen that not just a single but a whole generation of calves who have the same bull sire—that is, they are half brothers and half sisters—will have some obnoxious traits.

As careful he is, Richard can have a minor outbreak of wildness. Like so many things that can go wrong, it was both unexpected and also, once discovered, entirely explainable. He bought a very nice bull, nice in all respects. He was tall, long, broad where it mattered, and very gentle and manageable. And so were all his sons in the first calf crop he sired. It was the daughters. They didn't like people—they didn't like their own calves particularly. They were effectively protective in the sense that you couldn't get anywhere near one when it had a calf on the ground. Somehow that bull had acquired a sex-linked characteristic that could not be predicted until it expressed itself in the female offspring. This nastiness wouldn't necessarily matter on a beef-production program where range cows are left to their own devices, but it had to be rooted out in a breeding-stock ranch. In the very competitive business of selling registered Aberdeen Angus, a ranch has to be very good at the details, including selling only animals that can be managed. And the only way to fix the problem after these ringy heifers showed up was to withdraw all the bull calves (who might be carriers of the trait) and all the heifers sired by that bull from sale as breeding or commercial stock.

A breeder doesn't have to memorize and recognize all the different combinations of bull and cow on the ranch, although it is amazing how well a breeder does know most of upward of a thousand cows and calves. The ear tags on the animals amount to a sufficiency of information. Their own large numbers indicate their birth order, among other things, and smaller numbers and letters above the individual's identification point back to dam and sire. These latter numbers are more for convenience than anything, as each calf is given a number and entered in the herd book within a few hours of its birth. With the Angus Association breed book in hand, Richard could not only identify the father of a batch of obnoxious heifers but could run the ancestry of all the mothers and that bull back into the nineteenth century.

This particular nasty-daughter-producing bull had one special and valuable trait that Richard wanted on the ranch: He produced smaller-than-average-sized calves but ones that grew quickly and caught up with the heaviest of their cohorts by the time they were weaned. This trait is important when breeding heifers because the smaller calves pop out relatively easily; once a heifer has had her first calf, she can usually handle a large calf. Several years after the troublesome bull (and all its offspring) were out of the Feltons' registered herd, Richard tried another bull that had all the traits needed for a bull to be put on the heifers. He was distantly related to the troublesome bull, five generations removed; approximately one-sixteenth of his genes might be traceable to the trouble-causing animal. The seller claimed that the bad-heifer trait hadn't expressed itself in the bull's immediate ancestors, and so Richard took a chance.

Along came a small crop of bad, nasty, grumpy, flaky heifers. They were so obnoxious that nine-tenths of them were sold off for beef as soon as they were weaned. Their brothers were altered and sold as steers. But two of the flaky sisterhood seemed calm and amiable, and so Richard kept them and bred them. Perhaps, he thought, they had escaped the genetic imperative. It turned out they had been hiding their ugly side. The pair of them behaved nicely right through pregnancy, and as it became obvious the time was near for the first one of the two to calve, she was put into a pen inside the barn. With heifers, there's a need to check them all regularly, including a few times during the night. The ones in labor, or near it, are taken indoors. Richard was away in the afternoon, so I pushed open the solid door to her pen to take a look at her, and she tried to smash her way through it. With the door slammed shut, she went across the pen and tried to knock her way through a barrier made out of 2-inch (5-centimeter) iron pipe and bent it permanently out of shape. Then she turned and faced the door. I could see through the small void between the door and the jamb that she was the picture of rage. Besides snorting, she had her ears flattened back, a bad sign if one more was needed.

She had her calf easily enough before the next bed-check, and Richard tricked her into moving out of her pen and into the one next to it by giving her the impression that if she came over to that pen, she would be able to whack the bejeezus out of him. Then he and I went in to put a tag in the calf's ear and give it a shot of antibiotics. The calf was goofy too. Ordinarily, if you get hold of one, step over it with one leg, and get your legs around its neck and hold its head up, it calms right down. Not this one. I ended up tackling it and sitting on it. Even then it thrashed around so much that it took Richard three tries to get a tag pinned through its ear. In short order, both mother and daughter were at the packing plant. Not incidentally, one of the signs of uncontrollable nasty behavior in the heifers already sold off for meat was that they were constantly butting each other and any other heifer calf that would stand up to them. By not accepting the dominance order, they were outlaws among their own kind.

This unfortunate consequence of accidentally bringing some rankness into a breeding-stock herd is hardly a one-time event in the history of Aberdeen-Angus. It can happen to the best of breeders. It is just what William M'Combie meant a century and a half ago when he wrote of a mistake he made: "Inexpressible mischief may be done by the introduction of wild blood into the herd, for it is sure to be inherited." The mischief at the Felton Ranch was hardly inexpressible, not with Richard's vocabulary. "You would think," he said, among other things, "that five generations would get rid of that crap."

It seems cruel in a world of superabundance, with meals guaranteed and water assured, that calm and well-behaved domesticated animals should insist on hierarchy, on dominance. But cattle are just at the edge of domestication, and Aberdeen-Angus a little closer to the primal state than some breeds. The rancher has to take the whole animal. If he wants a cow that will protect its calf (up to a point and when appropriate), a cow that will range widely in a summer pasture, exploring to the fence lines for particularly choice morsels of grass and brush and in springtime for the delectable flower stalks of the soap yuccas; if he needs a cow that can stand stoically next to her calf in a blizzard and not drift with the wind, then he has also to accept the bullying, the chain of dominance.

In winter, with hay spread out once or twice a day, an older cow that is slow to move or a lame cow walking with a limp will quickly fall to the bottom of the pecking order. It may need to be taken out of the herd altogether and fed and watered in isolation. In their large and limpid-eyed apparent good nature, her sisters will unintentionally starve her, probably not to death but to weakness. Not only will she have the last place at the table, but she will be slow to get there. Being bullied thin may cause miscarriages and certainly, if she is already nursing a calf, put it at risk. And it puts her in the condition that says to the rancher, as though she were wearing a sign, "Send me to the packing plant when this calf is weaned. I'm just plain wore out."

The ranch is not a cruel place, not nearly so cruel as that

state of nature where all life is brutish and short. But the sharp and dangerous edge of nature penetrates into the wintry fields and the summer pastures. This constant creation and maintenance of the herd's hierarchy among these gentle and caregiving mothers is the price of being cattle, just as surely as sorrow is the price of being human.

⇥12⇤

The Aberdeen-Angus Enter and Escape an Era of Undersized Cattle

THERE is more to conformity in the world of Aberdeen-Angus than the size, shape, and behavior of the cattle. It isn't just cows that have a herd mentality. Throughout the long history of the breed, there have been episodes of conformity on the part of breeders, including the earliest days when it was decided that black was right and everything else was wrong. For almost the entire nineteenth century there was an excessive interest in the pure fattened size of animals, not just steers meant for butchering but also in the breeding stock. Obesity was thought to be proof of early maturing to market size. But the most difficult period to understand was the twentieth century's fascination with what can best be described as diminutive, stumpy-legged, barrel-bodied, broad-headed, flat-nosed cattle in all the British beef breeds, not just the black ones, but also Herefords and Shorthorns.

By the time the Feltons got into the breeding-stock business in the 1960s, diminutive purebred Aberdeen-Angus had become the norm in the United States. In Great Britain the little ones began to be the desirable animal before World War II. The first great breeders of the nineteenth century had turned out some animals that went to market weighing well over a ton (over 900 kilograms), and routinely sent 1,500-pound (680-kilogram) steers to the shambles. Their successors, a hundred years later, were sending two-year-old steers to auction at 700 pounds (under 320 kilograms). Richard Felton rememembers years when a 900- to 1,000-pound (408- to 454-kilogram) Angus steer was "a monster."

Although some of the largest strains of Aberdeen-Angus in the world come from New Zealand today (and have for more than a decade), the Antipodes are hardly united. An Australian chatting on the almost endless flight from Sydney to Los Angeles, via Auckland, New Zealand, turned out to be a cattleman, a registered Shorthorn breeder. His other seatmate added that he was into Charolais. Both of them had a dim view of Australia's Aberdeen-Angus, which were apparently still on the small side in 2002. "At home," the Shorthorn man said, "we call them 'black sheep.'" That is undoubtedly a prejudiced opinion, but there certainly are small Aberdeen-Angus out there.

What began this stampede to the light end of the scales is difficult to reconstruct at seventy years' distance. The small-is-beautiful trend began shortly after World War I in Britain, became very noticeable there by the beginning of World War II, and swept the Aberdeen-Angus world in the 1950s and 1960s.

The fact that it started in Great Britain is the first clue. The British economy went south even before the worldwide depression of the 1930s, and it continued to sink as the war in Europe started. During the war, food rationing in Britain made a beefsteak, let alone a whole roast, a commodity unknown to the average household. Although meat rationing had ended immediately after the war in the United States and

Canada, it had continued in Great Britain until 1953. There are some very clear hints that the pocketbook concerns during the thirties followed by the scarcity of beef during the war years influenced the consumer's opinion of what made a good roast.

The Aberdeen-Angus Society's journal printed several letters during the 1930s and 1940s that, while they seemed very likely to have been written by the journal editors, did make the point that when one could afford only a few pounds of roast beef for Sunday dinner, the Aberdeen-Angus was the most desirable. Even a few pounds of roast would make a cut that was, as an alleged customer wrote, wide enough that it would "stand up and be carved properly." With the light-boned small animals, a parsimonious portion at least looked like a roast, gave the illusion that one was serving a proper cut of meat. The same argument was made for steaks. Cut from a small carcass, a six-ounce (170-gram) sirloin strip would be thick enough to grill properly; a similar weight from a large carcass would be something on the order of a quarter-inch (0.6 centimeters) thick. This British choice of size so governed the breed that Maurice Felton recalls that when he was a young man Aberdeen-Angus were promoted as "baby beef." And indeed they were. The new small animals went to market at two years of age weighing just 100 pounds (45 kilograms) more than an old-fashioned standard-sized calf that had just been weaned at seven months.

It should be repeated here that even the smallest Aberdeen-Angus bulls turned out large and extra-large sons and daughters when they were crossbred with a large and high-quality cow from another of the British beef breeds. When Raymond and Margaret Felton were running a diminutive Aberdeen-Angus bull with Shorthorn cows at their old Nine Mile ranch, they regularly brought the largest just-weaned calves to the market in Missoula. The problem for the registered Aberdeen-Angus breeder was simple: Not all of his product deserved the showring or would attract a buyer who saw them at the farm, and all the culled bull calves (turned into steers for the occasion) and heifers would have to go to the feedlots and the packing plants. Breeding stock gets sold by its merits, its conformation, pedigree, and reputation. Culled animals get sold by the pound.

What a bad economy and rationing began was finally turned into a breed-wide obsession when the strain of stumpy cattle began taking prizes at the major shows and, more important, setting auction records for price. What drove the prices up on these little bulls in Great Britain was the foreign market, particularly Argentina, where cattlemen were looking for a way to put some marbling on their tall, rangy, skinny beasts. In the reverse of the usual result when crossbred with other purebred animals, Aberdeen-Angus bulls put on the near-wild cattle of the Pampas produced smaller, more compact hybrids. And as the British consumer now regarded small carcasses as normal and desirable, the Argentines modified their herds in that direction to keep their share of the market. Nothing worked better on too-rangy South American cattle than one of the little Aberdeen-Angus bulls.

British Shorthorns, by the way, also got to be nearly as small as Aberdeen-Angus, as did many of the Hereford strains. Photographs of all the British beef breeds taken in the 1950s and 1960s were posed to emphasize the shortness of the legs. In almost every case, loose hay is piled around and under the animal so that his brisket nearly touches the straw.

On the scale of agriculture in Great Britain, where a hundred animals was a huge herd, raising runty cows was reasonably profitable. If you produced a championship bull, the

Elevate of Eastfield

rewards were enormous. A striking photograph of a champion bull from the 1958 Perth show illustrates the ideal animal of the era. The bull set a record for Aberdeen-Angus, 25,000 guineas (£26,250). The proud owner and the auctioneer are shaking hands, and their clasped hands, with their elbows low and at their sides, are a good six inches over the back of the bull. If you had one in your house right now, he'd have to look up to see a doorknob. Apparently it wasn't regarded as humorous at the time, but this little stub of a bull was named Elevate of Eastfield.

Richard Felton took a look at the photograph and winced. "You know," he said, "we used to have some 'Elevate' in the herd, so did everyone else. I've been trying to get rid of those small animals for thirty years, and you know what, once in a while one will pop up."

The small animals just didn't work as purebreds on the North American range. This is a place where size does matter. If you put a little bull out on a square-mile pasture, and his legs are half the length of a typical bull's, he'll effectively have to take twice as many steps, go, in terms of effort, twice as far. The other serious problem was economical, as mentioned in regard to selling culled animals for meat. A typical breeding-stock ranch will send anywhere from 35 to 40 percent of its animals to slaughter, after culling the calf crop and making steers out of the less-than-excellent bull calves. And there are always more heifers than the market will bear. There just wasn't and isn't any market for undersized animals headed for feedlots. But before the fad was over, it had taken control of the British herds and was widespread in America, even impinging on the High Plains in the American and Canadian West, where steaks should be big enough to cover a plate.

Some of the famous herds in Scotland just withered away in the 1950s and 1960s because their farmers refused to join in the parade to puniness. Mulbane Mains, near Cullen in Aberdeenshire, was typical: once prominent, suddenly shunned. The last good sales from Mulbane Mains were bulls exported to the Wye Plantation in Maryland, a well-known breeder of bulky standard-sized Aberdeen-Angus. When the herdsman at Wye was traveling to Ireland and Great Britain buying breeding stock, he found that the animals he wanted—large, tall, and long—were always bargains when he could find one at auction. Most of the time he bought them off the farm—they just weren't up to auctioning. But Wye was an exception, and many American breeders looked at the auction prices in Great Britain and caught the baby beef bug. This all timed perfectly with, as noted in discussing dwarfism, the outbreak of dwarfing in Hereford and Aberdeen-Angus cattle that encouraged American ranchers to buy British stock.

The almost trite but very true concept of the swinging pendulum is the best description of what happened when the perigee of smallness was reached. Within a few years, hugeness was in, and the herd of breeders charged off in that direction. What gave one of Scotland's successful breeders, William McLaren of Netherton Farm, his head start in the 1980s was that he had never gone down the runt-road but carried on with medium-sized cows. That alone doesn't account for his recent prominence. Like most British breeders today, he improves his herd with genes imported from abroad, either as frozen semen or, to get a whole new line in the herd, as frozen embryos carrying new paternal and maternal genes. Due to some curious veterinary laws, it has been impossible to get semen from the United States and, for a while, from Canada, and the common import source now is through New Zealand. It may be largely (sometimes entirely) Canadian or United States bloodlines but quarantined, in a manner of speaking, by emigrating to New Zealand. You might think of it as "semen-laundering."

Rainy Brown, a longtime student of the herds in Great Britain, and the man who is expected to produce the next history of the Aberdeen-Angus Society of Great Britain, makes something of a hobby of keeping a close eye on herd genealogies. He may be the only person in Scotland, if not the world, who can identify from memory all the cattle in a show catalog that are carrying the recessive gene for red coats. This is not a matter of great consequence in Britain, where red cows are not discriminated against by the herd hook, but it is a sign of Rainy's encyclopedic approach. He remarked while watching an Aberdeen-Angus show and sale at Carlisle, England, that every animal there was at least three-quarters North American bloodline, and that most of them were from 96 to 100 percent non-British stock. Indeed, the classic British Aberdeen-Angus cow barely exists. One of the breeders showing at Carlisle (and showing two beautiful large but not grotesque bulls) actually keeps a separate herd of all-British animals. They won't win any prizes for him, but he has some of the spirit of old M'Combie in his effort to preserve the original blood.

How the breeders managed to shift from quarter-ton beasts in the 1960s to strains where bulls may reach a ton and a half is an interesting question. There is no doubt that the original Aberdeen-Angus carried genes that could be selected in favor of different sizes, just as they had genes that could allow selecting for different colors—all black, all red, and black or red with considerable white. And if the flexibility was there to turn out little things like Elevate of Eastfield, then it should be possible to revert back to the large animals that were desired for most of the breed's existence. But going back to the reasonably large animals was not enough. Breeders in Britain,

Aberdeen-Angus show and sale at Carlisle

parts of Canada, and New Zealand are trying to outdo the traditional, original Aberdeen-Angus. It almost looks as if some breeders took a shortcut and slipped in some outside-the-breed genes. One of the easiest ways to put some size on a strain of Aberdeen-Angus without getting caught out is to mix a little Holstein-Friesian into the gene pool. If you have ever hung around a dairy herd of those cows, you can't help but have noticed how huge around they are.

A now-retired cattle geneticist, completely familiar with the Aberdeen-Angus cattle of North America, answered a question obliquely but clearly (and not for attribution). Asked about the possibility that there was a bit of Holstein in some of the Aberdeen-Angus seen in some herds today, he equivocated but just for a moment. When pressed just a little, especially about the curiously high pin and hip bones on some Aberdeen-Angus cows, contrasting with the well-covered bones on a rounded-off animal, he said this: "Did you know," he said, "that the only two breeds you can't tell apart with standard blood tests are Angus and Holstein?

"It's always been that way," he continued. "I think there were two times when Holstein came into the breed; one was a very long time ago, and one was very recent. By the way," he added, "Holsteins marble up just like Angus."

Just a few winters ago, one of the cows on the ranch decided to get a drink of water out of a ditch by the road instead of walking another fifty yards to a water tank. She slipped, went through the ice, and ended up stuck in the ditch like a spun-out sedan. She was one big old cow and couldn't be dragged out of the ditch with a rope and the pickup. Maurice and Raymond had to use the big bidirectional tractor like a wrecker, slide her up on a flatbed trailer, and haul her off to the barn by Richard's house. The oddest thing about the cow, and it says something about the will to live in cattle, is that when thrown some hay she could reach, while everyone waited for Maurice to come with the big tractor, she started eating. Once in the barn she kept right on chewing. Here is an animal on the verge of death, and somehow it senses that a square meal is a useful medical treatment. When she got water in a low-sided container placed right next to her head, she took a drink. To Richard's amazement, when he stopped back, she was up on her feet apparently recovered.

This particular cow wasn't part of the Felton Angus Ranch herd; it didn't have the lazy Y K brand. It was not a typical Felton Angus Ranch animal. Its main and maybe its only good point was hugeness. It belonged to one of the boys, one of Margaret's grandsons. When told we'd gotten the cow out and it looked like it might live (and have its calf), she was pleased but hardly overjoyed. "I don't know why he bought that thing in the first place," she said. "He spent a lot of money on a big old cow that looks way too Holstein to me." It survived and had a calf. The calf didn't make it to the production sale; it just didn't thrive.

There's another clue to some Holstein blood in a beef pasture. Every once in a while at calving season, very rarely in a herd as thoroughly purebred as the Feltons', but it has happened there, out will pop one of those black-and-white things that look remarkably like what they are, Holsteins but hornless. When a genetic defect called mule-foot appeared in a few Aberdeen-Angus strains (thankfully not in the Tongue River livestock), some cattlemen suspected it was another sign of Holstein blood, an indication of how widespread was the sense that something was being fiddled with. The name mule-foot derives from the fact that it causes one or more of the foot bones of the victim to fuse together, and what was once a cloven-hoofed animal then leaves a footprint like that of an equine.

If all this did was make the animal look odd, mule-foot wouldn't be a problem, but it is the outward sign of another defect, a basic problem in metabolism. The mule-foot animal can't regulate its body temperature properly. When it is under stress, it will die of heatstroke at the first opportunity. Mule-foot is properly called syndactyly (bio-Latin for "fused digits"), and it is described in veterinary texts as a rare disease of Holstein-Friesian cattle. It has appeared suddenly in other breeds, but it was unknown in Aberdeen-Angus until late in the twentieth century. It is understandable that some cattlemen went for the simplest explanation, which is almost certainly not the correct one. Mule-foot in Angus was traced back to just two bulls (a father and son) whose offspring, when they bred together, turned out the usual frequency of genetic problems. A quarter of the offspring showed the disease; on testing, about half carried just one copy of the mule-foot gene and were outwardly normal, and the last quarter didn't carry the gene. In those few Aberdeen-Angus cattle that expressed the genetic flaw, they did so in a manner quite different from mule-foot in Holsteins. In Aberdeen-Angus, it often involved both rear hooves and not infrequently all four. In Holsteins, it most commonly involves only the right front foot, rarely that one plus the left front hoof, and even more rarely, one or both of the hind hooves. It was fairly simple to eliminate this strain of mule-foot from registered Aberdeen-Angus herds, although financially painful for a few breeders. Unlike the episodes of dwarfism, there was no search for environmental causes, no demands for a cover-up.

You might wonder how a lethal genetic disease manages to survive in the Holstein gene pool if it keeps killing its hosts. As long as the evil gene is just sitting there on one side of the spiral helix of chromosomes not bothering anybody, the animal is fine. However, it should be rarer than it is in Holstein cattle, being as lethal as it is when it occurs. There is evidence that dairy cows who carry a single copy of the gene for syndactyly/mule-foot also generate higher butterfat yields. Without knowing it, the dairyman may have selected animals in a way that favors the destructive gene. This is the old bugbear pleiotropy, or the ability of a single gene to affect more than one thing in the organism.

Another odd thing is happening to Aberdeen-Angus, although not in the United States, where the American Angus Association forbids the registration of an animal with this new and troubling trait. This is the condition popularly called double muscling, which gives a beef animal the market advantage of having huge leg, rump, and shoulder muscles. They aren't double muscles; they're just the same old number of muscles all grown huge. The right word is hypertrophy. This is the most extreme form of breeding for leg muscle, and it is not clear whether some breeders are capitalizing on normal variation or whether some crossbreeding is under way.

Hypertrophy is the norm in a breed called Belgian Blue. In countries with a prejudice for lean meat, Britain, Australia—and in some regional markets, including the West Coast of the United States—the easiest way to get gobs of mature lean beef (like round steak) on an animal is to raise (and crossbreed) the Continental oxen for their rump and round and stew, to use butcher's terms. When hypertrophy shows up on a registered Aberdeen-Angus in Scotland, you have to wonder if someone, somewhere, didn't add in a touch of Belgian Blue or another Continental cow. Not that an honest breeder would do it, but the way semen and embryos are flying around the world, what looked to a buyer like just a nice heavy-shouldered Aberdeen-Angus strain might have been an adulterated or improved strain.

Like so many "improvements" on the model Aberdeen-Angus body, hypertrophy is associated with any number of problems wherever it appears. It is this group of related problems that matters more than the distribution of muscle, and it is the primary reason for the American Angus Association's refusal to register hypertrophic stock. In Belgian Blues, where hypertrophy is almost universal, herdsmen routinely have to pull calves or perform cesarean sections: The calves are too big and the double muscling can constrict the birth canal. Even after the calf is out and grown up, it is typical of hypertrophic cattle to have all the extra muscle, the extra volume, but never to have a larger basic support system for all that extra flesh. Their lungs and their hearts are merely average. Imagine a weight-lifting human being who has bulked up to 300 pounds (136 kilograms) with insufficient cardiovascular development. Like some professional American football linemen, hypertrophic cattle have no cardiovascular reserve and become exhausted easily when exercised. Very large linemen are occasionally struck dead with cardiac problems. In North American range conditions, double muscling is a terrible fault, however beneficial it might be in small, closely watched European or New Zealand herds.

It may be too late for the British society to ban the registration of the hypertrophic animals that are starting to appear on British farms. It's much easier in all situations to stop something before it gets started than to take it away later. If it does become part of the British herds, the consumer dedicated to lean red meat should feel no anxiety. This is not a matter of interspecific "genetically modified foods"; this is not like putting beef genes in a chicken (which might give it the proverbially scarce teeth).

This is intraspecies modification by mixing compatible genes, something human beings do all the time themselves.

Even in New Zealand, the source of several strains of oversized Aberdeen-Angus, the large animals are not universal favorites. While I was visiting New Zealand's South Island, I fell into conversation with a farmer in the very small town of Whataroa. He turned out to be an Aberdeen-Angus man, and we chatted about the breed and some of the trends. Without prompting, he remarked that he was not at all impressed with the new large strains. "I see them in the paddock," he said, "and they do not forage. They stand around."

Besides any intrabreed finagling or, more likely, just the mindless pursuit of spontaneous hypertrophic variation within a single breed, there is something going on that can best be described as brutalization. Cattle are by definition brutes in one common meaning of that word, but there are more refined and less refined cows, and Aberdeen-Angus can be quite refined. But in the search for size, the appreciation of symmetry has diminished.

A group of visitors and Aberdeen-Angus breeders and association officials were watching the veterinarian check on the animals just prior to a show and sale in England. I suggested to Rainy Brown, the genetics whiz, that there should be a friendly contest to pick which animal would be judged the champion bull as he came through to be vetted (the original meaning of the word we now use for checking up on human beings or their intellectual output). And so we did. I picked a handsome bull, but Rainy scoffed and said it couldn't win. As a particularly large and lumpy bull came into the squeeze chute, Rainy opined that this was the one that would take the prize. In fact, largely by beginner's luck, I had picked the champion, as far as agreeing with the official judge meant that I had gotten it right.

That is, it was by far the best bull under the judging standards of the Aberdeen-Angus Society of Great Britain and was so noted and rewarded by the judge. But when sale time came, the grand champion brought the fourth-best final bid, several hundred pounds less than Rainy Brown's lumpy choice. The second and third bulls who took top money had no special virtue except size. "I told you mine would be the *real* champion"—Rainey chortled, not *too* gleefully—"and I say now what [Queen Victoria's Prime Minister] Disraeli always said, 'Let the people decide.'"

Up in Alford in Aberdeenshire, across the valley from M'Combies' farm at Tillyfour, they have put up a wonderful bronze sculpture (artistically speaking) of an Aberdeen-Angus bull to commemorate the establishment of the breed. It is life-size and hyperrealistic. The artist sculpted it after taking hundreds of measurements with calipers and rulers and photographs from all conceivable angles. They picked a logical enough animal for the model, a Royal Highland Show grand champion bull. And he's big. And he's kinda ugly. The Felton bulls—and the same could be said of most registered bulls from the northern plains—are better examples by practical standards. Bill Sklater, the Orcadian judge of Aberdeen-Angus, had not seen the statue before. He looked at it for a long time and said: "The animal's got a bad fault. Look at those legs."

The bull was so barrel-chested that his front legs were bent. The only way they could get back under his chest and line up with the hind legs was to be bent inward at the hock. The animal was knock-kneed (or knock-hocked, to be anatomically accurate). They're probably perfectly good legs for a small paddock in Great Britain, but that dog can't hunt, as the saying goes, not on the High Plains. The animal was not made for walking. The other problem with the statue is that it has minimally sized testicles. Another centimeter or two smaller in diameter, and he wouldn't pass the veterinary inspection for showing in a breed championship competition. Of course, it was much easier to squat and measure the bronze scrotum than it was for the artist to put calipers on the real thing.

One of the lovely things to like about Scotland is that they still have an agricultural way of life; even townspeople know something about animals. It's the kind of country where a tourist can get down on his hands and knees at roadside and measure a bronze bull's scrotum with his thumbs and forefingers without attacting attention from the passing drivers. The bull had a marginally satisfactory set of testicles.

. . .

Small articles appear here and there in such publications as the newsletters of auction houses and local agricultural society bulletins with headlines like ANGUS BREED MUST RETAIN EASY CALVING ADVANTAGE. What they are talking about are bull lines that turn out too-large calves and cows with too-big butts that impinge on the lower end of the reproductive tract, even without hypertrophy. The Aberdeen Angus normally has the trait of turning out medium or smallish calves that grow rapidly and catch up with animals that started out life weighing more. "It has been noticeable," a farmer wrote in the Perth Agricultural Centre auctions magazine in 2002, "that many of the most recent Angus cattle imported from Canada are sired by the three most difficult calving bulls in North America." He noted further that American range cattlemen, who've had big-enough bulls for years, were there with large-animal strains "when the high-flying pedigree cattlemen needed rescuing [from the fad of stubby cattle]." He reminded the readers that Great Plains breeders had provided Britain with satisfactory bulls that turned out reasonably sized calves. Referring back to the difficult calving Canadian bulls, he concluded that British farmers shouldn't "touch these bloodlines." But from the looks of show animals, a number of Continental and British breeders are embracing the big bulls and taking their chances with oversized calves and undersized birth canals. A skilled herdsman with a few confined animals to tend can overcome most difficulties in births.

That is not to say that Aberdeen-Angus in general, and the Feltons' animals as well, are obstetrical wonders. Pulling the occasional calf that's coming out backward, for example, is usually a necessity. Sometimes if the rear hooves are outside the cow's body, things will succeed naturally. But a pure breech presentation with the hind legs tucked up under the calf's body is something no cow can solve by herself. Recently, one Felton calf presented as a breech birth and required considerable manipulation. The water had broken and the only thing visible was the poor thing's tail. Richard had to reach up into the uterus, get a hold on one of the hind legs, and pull it down and out, then get the other leg out. This can be difficult—things are a little slippery up in there. Then the ends of a short piece of fine-linked chain are looped over each hind hoof and the calf is pulled. Someone seeing a calf pulled for the first time would swear the legs were going to rip off, but the calves are well constructed.

When the calf did come out, it hit the ground with a thump and didn't even quiver, let alone move its head. We gave the calf CPR after it lost its heartbeat, cleared the mucus from its airway, breathed for it, revived it, and then lifted it up by the heels with the pulling chain looped over the bale-lifter on the pickup to let any aspirated mucus and amniotic fluid drain. It survived. Richard likes to refer to it as "the calf we brought back to life." It deserves to be sold as breeding stock and have a nice bull's life, having made such a narrow escape from no life at all. It doesn't pay to get sentimental about cattle, but that doesn't mean it's wrong to root for a tough little surviver. Unfortunately, the mother died, and the calf couldn't bond with another cow. Hand-fed, it didn't thrive and joined the steer herd.

Too-frequent calving problems, something Richard avoids like the plague, are an example of what he keeps saying: You push any trait too far in one direction, and a problem will pop up in another place. That is a western cattleman's understanding of the way the real world works, not peculiar to the breeding program at the Felton Ranch. That is exactly what the Scot farmer meant when he wrote of the debt owed by the British stockmen—not to some fancy breeder but to a whole class of North American ranchers. It was nice to read a rare good word about the "range cattlemen," who get very little press in the publications dominated by, as the Scotsman said, the "high-flying" breeders. Going to extremes gets the publicity; holding the center wins the battles.

The Felton Angus Ranch is one of many classic range operations in the American and Canadian West, not a college experimental station or a boutique breeding farm. When fads come and go, when fashions change, it is good for an industry and a breed of animals to have a center of gravity. At this point, the center is being held together by North American breeders on the extensive ranches where the animals must have some gumption and some self-sufficiency. The hard-birthing Canadian stock referred to by the Scot in the Perth auction newsletter are not what visitors see on extensive ranches in the Canadian West.

From the very earliest writing about Aberdeen and Angus cattle, they were noted for three virtues. First, they were amenable and amicable, easy to handle. Second, they were good at converting a rough and spare diet into quality beef in a reasonable number of months. And third, they were capable animals; they did not require close supervision; they were good foragers, easy calvers, attentive mothers. A few of these qualities can be dispensed with in small, closely held herds. If one can tame a tiger, any cow (excepting dairy bulls) can be gentled. If rich food can be had at a reasonable price, any breed or any particular strain within a breed can be fattened. And if skilled herdsmen and veterinarians are on hand, calves can be pulled if possible or delivered by cesarean surgery if necessary. Rejecting mothers can be haltered and penned with their calves; if that fails, the cow equivalent of a wet nurse is as close as the nearest dairy if not already in the herd as a cow that has just lost its own calf.

For all the difference in landscape and climate and scale, the Scottish farmer of the nineteenth century prized and bred for the same qualities as a range cattleman on the High Plains does today, with the sole exception that the original animals had to have the temperament to put up with close confinement. Fads will arrive and fads will disappear, but somewhere, probably on the High Plains, as the Scotsman wrote, the range cattlemen will hold the center for the breed. As long as there are cattle in the Western states and provinces, the original, the historical Aberdeen-Angus will survive and thrive. The breed has a new and natural homeland.

Epilogue

THE Aberdeen-Angus cow and its kindred have made an enormous journey. It began with the most mysterious domestication human beings ever achieved when they diminished in size a great and terrible animal, dulled its ferocity, and then taught it to pull the plow and the wagon, give milk on demand, and produce meat that for carnivores is the standard of perfection. So many odd and unfamiliar foods are said to taste like chicken. Nothing else tastes like beef, and some believe that no beef tastes as good as properly handled Aberdeen-Angus. A few other breeds have their devotees, and they are welcome to disagree. The argument, after all, is just about which is the greatest of beef in particular—an intramural fight, which everyone wins, for in the end everyone in the battle is eating the best of all possible animals.

Despite the occasional irrational trends and fads that occur in the cow business, the possibility of a new race of cattle so superior to our current models that established beef and dairy breeds will disappear from the earth is quite remote. Ten thousand years of animal husbandry (give or take an eon or two) has produced an amazing variety of cows, nearly a thousand different breeds, mostly local curiosities. But for all their variety, there has been no fundamental change in the original model, just exaggerations of one trait or another at the expense of a third trait. There is no great all-purpose cow, but there are great cows for all purposes.

As in all human endeavors, there are people content to maintain a high standard with the historic products of civilization (no one has thought of a better way to make a violin in the last three hundred years since Amati and Stradivari) and there will be people who want to push the limits with, for example, electronic devices that can pretend to be a violin. So it is that cattle breeders will keep tinkering, trying to achieve an imagined triumph. A good deal of the interest today is in identifying genes that will tweak some commercially useful trait. At least two groups of researchers, one in Texas, another in Montana, are looking for a gene (or genes) for flavor and a gene for tenderness. Those are not well-defined terms. One person's easy-to-chew is another's mushy; one gourmet's flavorful is another's gamy. At the Montana research station, in addition to the laboratory genetic investigations, herdsmen are trying to achieve a great goal by old-fashioned crossbreeding. Most recently, they were crossing Japanese Wagyu with Charolais, hoping to inject some extra tenderness and flavor into that big blond French cow. People familiar with Aberdeen-Angus cattle are not at all sure that this is necessary. They think the job was done around the time that Amati and Stradivari perfected the bowed string instrument. There is no secret to tasty tender beef. Any chef can tell you: It must be bred well, fed well, killed well, hung well, butchered well, and served rare.

Aside from getting rid of horns on several breeds, centuries of selection have made just a tentative start at one other fundamental structural change. Beef animals are starting to appear with an extra false rib (as ribs not attached to the sternum are called). This makes a fourteenth rib, but it is not arising consistently; it is apparently a fugitive and recessive trait and is not as fixed in any race of cattle as is hornlessness. Another inch of rib steak may not be worth the effort to select for an extra bone in a strain of cattle. It will become easier to do something like adding a bone as science progresses—whether it will be wise is the question. The possibility exists now, and will only become more possible as time goes by, to alter the very genes in cattle, to remove this and replace it with that, but it is hard to imagine what can be fundamentally changed, as opposed to improved, in a way that would create a new breed that still could successfully live in the rugged conditions of the High Plains, thrive indoors in a damp Scottish winter, and emigrate successfully to the entire temperate world both north and south of the equatorial tropics. And taste good everywhere.

Even in the most advanced future society we can imagine, we may owe cows for a whole new class of benefits to mankind as genetic engineers succeed in getting cows to produce, for just one example, organic pharmaceuticals in their milk. The current mammal of choice is the goat for such transgenic work; goats are a convenient size and can be handled by technicians with a minimum of agrarian expertise. People are entitled to have qualms about transgenic animals, say a pig with a human immune system that would provide an unlimited supply of organs and tissue for transplants into people. This could also create a new source for diseases once limited to either species, now able to infect the previously immune animal (including the human animal). But if animals do become transgenic pharmeutical factories, I for one won't mind seeing cows step to the forefront of medical progress. When it comes time to make medicine-producing milk on a grand scale, milk full of something odd and useful, it should be from cows. It would be nice to see one more link on the long chain of our mutual lives, one more example of the association of cows with human progress—from the Pottery Neolithic right through into the era of biotechnology.

. . .

How long there will be Aberdeen-Angus in Montana specifically will depend on the new owners who are taking up more and more of the rangeland. For the most part, people buying very large ranches for privacy and recreation are keeping the cows. Advertisements for extensive cattle ranches with a recreational potential almost always note that "management is in place." There are any number of good reasons, none of them altruistic, for running cattle. The ranch operation can help with taxes (there's no law against losing money as long as you're at least trying). Newcomers to ranches with irrigated fields will certainly want to keep them in shape. There's nothing like agriculture to attract game birds and provide a level and nutritious landscape where the deer and the antelope play. And while cattle will not make you wealthy, they can make you look wealthy and feel wealthy. The oldest Latin word for fixed assets is literally "wealth in cattle," *pecunia* or *peculium,* mother-word for our "pecuniary," as in interests. Cattle make us feel rich; it is bred in our bone. The new landowners have a choice: a "second home in Montana" or a "ranch in Montana." The latter has undeniably more cachet. As happened with so many breeds of cattle in nineteenth-century Britain, it may be the landed rich who preserve them in the recreationally attractive American West in the twenty-first century.

In Scotland as well, considerable property is being acquired by out-of-country money (and that would include English money). With a wry good humor and a nod to Britain's colonial past, these new landlords are called "white settlers" by Scots. Newcomers there, just as in the American West, are buying estates for privacy and hunting grounds. And a few of them have gotten serious about cattle. Men with distinctly non-Scot, even non-European names are bidding again at Aberdeen-Angus auctions, but this time they are not foreign buyers—they are neighbors. There aren't enough examples to make a case, but national origin seems to make a difference in how the white settlers use their new property. Some immigrant owners from the Middle East and Asia, both aboriginal cattle-domestication areas, are keeping cattle on their estates and appearing at shows and auctions. Visitors at a cattle show will hear a man who speaks English with an otherwise heavy accent talk with pride about his "coos" as if he were to the manner born. He had Aberdeen-Angus coos and excellent ones. Oil-rich newcomers from the Arabian peninsula appear to stick to the horsemanship and falconry of their native culture, but no clear trend is established.

From the day the first Neolithic woman (speculation, yes, but a woman is the likeliest candidate to think of something to do with a cow besides eat it) took her pottery jar and squeezed some milk into it, we have lived symbiotically with cows. Oh, sheep and goats are fine milkers too, but try to get a goat to pull a plow or get a sheep to do anything positive besides staying away from the sheep dog. Even if it weren't more stubborn than a mule, a sheep wouldn't be much good at pulling a load, and the thought of a team of sheep is beyond ridiculous. Simply put, we owe cows an enormous debt for all the food and labor they have given us. We do not have to worship them like the Hindus, but it is sweet and proper to make room for them in our world.

Cows are also wonderful just to look upon. Managed sensibly, they do no harm, and in some parts of the world, they are a great aesthetic asset. In the state of Vermont, and the same could be said for Scotland, the cows themselves and the men who raise them have preserved a scenery without equal, not as landscape architects but as utilitarian managers. Most of New England is either overgrown with brambles and forest or carved up into house lots, sometimes both at the same site. Vermont, with a stubborn resistance to a mindless marketplace, has made every effort to keep cows on the farms. There are cooperative dairymen making cheese of national renown and even a nationally distributed ice cream manufacturer dedicated to buying local milk. Additionally, New England dairymen get price supports that keep small and less-profitable farms in operation. The result is openings in the otherwise oppressive vegetation, patches of pasture high above the valley floors that by their very absence of trees make the adjoining woodlands all the prettier. They're called high mowings in Vermont, and the Green Mountains that gave the state its name would be much the drearier without them. In the ancestral home of the Aberdeen-Angus, the neat fields and paddocks that edge up to the moors have a similar effect. They provide a sort of visual boundary to the civilized world, before the eye enters the less tamed vista of heather and gorse. And the cows themselves, well, cows make people smile.

It is pretty to think, à la Henry David Thoreau, that in wildness is the preservation of the world. But if we like human beings and enjoy the fruits of civilization, we must acknowledge that in the act of domesticating cattle and the attendant development of agriculture, our ancestors found the material success to become people who have complex ideas like wilderness and deliverance—and the language to write them down.

Acknowledgments

THIS book simply would have been impossible to write without considerable hands-on assistance. I am greatly in debt, from the first day to nearly the last, to the reference and circulation librarians at the Concord Free Public Library in Concord, Massachusetts. They never failed to find the obscure books I needed, reaching as far as the University of North Dakota in one direction, and the Universities of Maine and Texas in others. For this herculean work, there was a price: twenty-five cents a book. The librarians of the Tufts Veterinary School in Grafton, Massachusetts, seemed to enjoy helping a visitor who was openly and entirely ignorant of the indexes to the technical literature and of their catalog system.

Books are one thing. Reading cattle is another; for that, you need cows and a ranch. My cousins Margaret and Raymond Felton were good to me when I was a boy, and when I descended on them at their ranch after a gap of some fifty years they appeared to have been expecting me any day. Their sons and daughters-in-law, Richard and Karol Felton and Maurice and

Kathy Felton, who also live on the Tongue River ranch, opened their doors without any questions beyond did I want bacon or sausage for breakfast and ice cream on my strawberries. When my neighbor and illustrator, Gerry Foster, and his son Mark appeared they were warmly greeted and given a thorough introduction to the various animals on the ranch.

In Scotland, three members of the Aberdeen-Angus Society of Great Britain, Willie McLaren, Bill Sklater, and Rainy Brown, fed me, drove me all over the countryside, introduced me to a host of cattle farmers, and tried to teach me some of the finer points of the breed. It was an astonishing display of courtesy, kindness, and good humor. Scots in general were helpful, chatty, and welcoming, which did not entirely surprise me. They had treated my father just that way when he came to the country in 1941 to oversee the construction of much-needed airfields and naval bases.

For some twenty-five years, beginning when he became the editor who signed my first book contract and subsequently became my agent, Richard P. McDonough has been unfailingly supportive. He has not only brought me contracts (that's the easy part) but also suffered through the rough spots most loyally.

My wife, Florence, has had more conversations about cows than she really needed to, although she is generally pro-cow, and put up with my frequent absences in my quest for yet one more idea about, one more image of, one more insight into cattle. This is just a short list of helpers and supporters, but my gratitude should not seem diluted by being spread thinly across too many names.

Appendix
Urban Myths and the Rural Cow

THERE is considerable misinformation circulating both as oral urban myth and Internet chat that can keep people from appreciating cattle for themselves, not for what some imagine them to be. Scotland is entirely free of misinformation about cattle. There are just too many people who actually know about cows for tall tales to thrive. The myths are all of recent North American invention and confined to urban and suburban communities distant from cattle country. Persons raised in Miles City, Montana, for instance, have never heard of such things.

Upon hearing that the author was about to begin a book on cattle, and the brand-familiar Aberdeen-Angus in particular, several people wanted to know if the book would include a section on cow-tipping. These were mature, intelligent, employed, and otherwise responsible citizens. Cow-tipping, many Americans believe, is a recreational sport of fraternity boys and other boisterous youth. As the story goes, cows sleep standing up, and if several hearty louts sneak up on one side of the cow and push it smartly, the cow tips over. This is particularly amusing to callow youths because the tipped cow is unable to rise, and until they are driven off the farm by a pitchfork-wielding furious farmer, they can stand and point at the cow and laugh hysterically. It should be noted immediately that cows lie down without being tipped and get up without assistance. Where the idea came from that they sleep while standing is mysterious, but it may have come from too much time spent watching Holstein-Friesian dairy cows who don't do much and look like they're unconscious on their feet.

The original source is an enigma, but the popularity of the cow-tipping myth thrives owing to several movies about undergraduate amusements (*Tommy Boy*, for example) in which cow-tipping is a highlight. For a while it was as mandatory a scene for college comedies as, say, driving into the sidewalk fruit stand (Watch those cantaloupes go!) became for car-chase sequences.

Some people believe it is true because they know someone who knows someone who tipped a cow. And then there are even more persuasive sources. Just last year an author of my acquaintance who was writing a book of unusual animal facts announced that she was including an entry on cow-tipping. When I suggested that this was not an authentic animal fact, that it didn't even rise to the level of factoid, she bristled visibly (her hair didn't exactly stand on end, but she looked as if it might any minute) and announced that she had "sources." Her main source, as she related it, was the top from a bottle of an energy drink; there, printed on the inside of the cap, was a brief and persuasive account of cow-tipping. At her request the bottler had sent her a list of all the animal facts it prints inside the caps.

One of the difficulties with cow-tipping, if ever attempted, is that cows weigh upward of a ton (907 kilograms). Assuming they are balanced on their four hooves, the two hooves on the cow-tippers' side are holding up as much as 1,000 pounds (453 kilograms), which may require more cow-tippers to lift and push on the beast than can comfortably fit onto one side of a cow. And the other problem is that cows are kind of soft and their hide is rather loose. It is hard to get a good push on a soft, flexible, movable surface. That is why all weight-lifting and weight-tossing events use instruments like steel bars or metal balls (or tree trunks, in Scotland), things made out of stiff materials that allow a proper grip and make a firm surface on which to push or pull. But it is hard to prove a negative. If someone swears that he actually helped tip over a cow, what can be done? Good advice would be to watch one's wallet.

The same bottling company that certified cow-tipping has also printed a cap with the interesting animal factoid that cows sleep with their eyes open to watch for predators. One of the beautiful things about the human brain is its capacity to hold two contrary thoughts without a meltdown. (Incompatible thoughts may create a psychological state called cognitive dissonance, which makes people uncomfortable, but they survive it.) One might ponder this: If a cow has its eyes open while asleep, how can someone sneak up on it and tip it over? Really, if it's light enough to see the cow . . . Additionally, one might even wonder how the cow can see anything while it is asleep. Men who snore in their sleep cannot imagine what the bed partner is talking about. "Me, snore?" Sleep is the absence of physical sensation, including hearing, seeing, and touching. This interesting and also true animal fact about sleep somehow is not applicable to cows, if one believes in bottle caps. There are no myths, urban or otherwise, in which any other beast sleeps with its eyes open and watching. Cattle are thought by city dwellers to be living in a parallel universe unaffected by ordinary mammalian physiology.

.　.　.

An actual interesting animal fact about how cows see (with their eyes open while awake) doesn't lend itself to being squeezed onto the inside of a bottle cap. It is more than intriguing; it is a remarkable example of evolution's role in shaping the various species. In a very real sense, cows do not have peripheral vision like human beings (or elephants, for that matter). They have something much better.

Humans have evolved to do a lot of fine work right in front of their heads. Even making the simplest flaked-stone tool requires extremely good central close-up vision, and eye-hand coordination to match. Stone-tool-making cultures are considerably older than modern man; the earliest hominids have left a scatter of chipped rocks formed into crude instruments. Good central vision was required because of the way the first human stone tools were manufactured. These precursors of modern man had to bash one stone very hard against another stone held in their other hand. Even the simplest tools require aiming at a specific spot on a rock held in one hand and whacking it right on target. Missing and subsequent finger bashing gets mankind nowhere. The fact that the structure, the organization of elephant eyes, resembles that of human eyes seems odd at first, but the elephant with his prehensile and sensitive trunk also does a lot of fine work right in front of his head.

The way humans and elephants achieve this very accurate frontal vision is by concentrating most of the vision-producing cells in a central spot in the back of the eyeball. The rest of the light and color reception equipment is thinly scattered across the remainder of the retina. The central spot is the macula, just doctor-Latin for "spot" (and easy to remember by the commoner English word "immaculate," e.g., spotless). When macular degeneration sets in, central vision is lost and peripheral vision will eventually be all the person has. Human peripheral vision is lousy. Compared to 20/20 vision with the macula, peripheral vision runs from somewhere around 20/200 to 20/400. Our peripheral vision can barely make out at twenty feet what the central vision of a normal eye can make out at two hundred to four hundred feet. Put another way, if someone has only peripheral vision, he or she is legally blind. There is a simple test that quickly gives a real insight into what it means to have only peripheral vision.

Here is the test, which a reader can perform without any special equipment. Looking straight ahead, put one hand out to the side with the arm straight and keep looking directly forward. With the hand then at ninety degrees to the line of sight and without moving the eye, try to see the hand. Most people will not be capable of doing this. A few more will be able to see the hand at that angle if they wiggle their fingers. Most people will have to move their hand toward the center before it comes into view, and they will get their earliest glimpse if they keep wiggling their fingers. The human eye and brain are, like all other animal vision systems, very sensitive to movement. As soon as the hand comes into view, stop it right there and also stop wiggling your fingers. The next and last part of the test requires a temporary suspension of belief. Pretend that you do not know how many fingers you have: If you pretend to count, to actually see, the individual immobile fingers on that hand at the point where you first saw them wiggling, it will be impossible. That is how bad peripheral vision really is. Most people will have to move their hand several inches (or degrees) toward the center before the individual unmoving fingers come into focus. Humans are poorly equipped to detect a lion sneaking up on them from the side if it goes very, very slowly (as lions do on television). It will be close enough to eat them before they see it unless a careless motion attracts attention. If lions would quit switching their tails back and forth, there would be more lions and fewer people. When a mystery novel author writes, "Out of the corner of her eye, Prunella Hawthumble saw something move . . ." he is talking about the widest possible field of peripheral vision.

In the sense just described, the cow lacks peripheral vision. The cow just has excellent vision out to the periphery. The cow's vision equipment (the rods and cones and accessory nerves) are not in a central spot like our macula but in a streak that runs horizontally across the entire retina. The cow sees just as well what is happening on the horizon to her left and right as she sees what is happening ahead. Other grazing animals of the world have the same visual streak, including the red deer of Scotland and its cousin, the American elk. The assumption, the presumed explanation, is that grazing animals are creatures of woodland glades and open savannahs and need to know what's coming up over the horizon or sneaking out of the woods toward them (a lion, say). This is more important than very precise vision in front of their noses. Elephants graze too and have peripheral vision as bad as humans', but what in its right mind or normal behavior tries to sneak up on an elephant?

It may have occurred to the reader that this is all very well when the cow is standing up with its head parallel to the ground, but what about when it is eating grass, certainly a major occupation for cattle? This would turn its excellent peripheral vision into a good view of the ground beside its nose and the sky above its head. The grazing animal's visual streak has an interesting adaptation that deals with the problem. At the front end of the retina (toward the nose), the streak develops a hump of concentrated vision cells that project straight up from the lateral streak at a ninety-degree angle. So when the cow (or the deer or the elk) bends its neck to snuffle up grass, this bump of vision cells on the streak points toward the horizon before it, and the animal maintains excellent vision ahead, although it does lose its complete horizon-to-horizon, back-to-front, view of its world for the moment. It is a compromise, but a good one. This same heightened forward vision allows the fighting bull to charge with its head down and its horns pointed forward, and on good days (from the bullfighter's perspective) cause the animal to follow the movement of the cape and ignore the matador's thigh.

·   ·   ·

It is difficult to compare the vision of a cow with a human being's or with that epitome of fine eyesight, the eagle's. Surely a cow's eyes cannot rival the eagle's, but compared to the eagle,

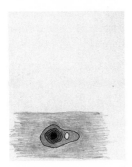 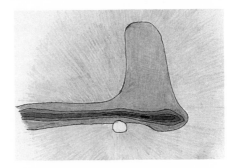

Diagram of the surface of the retina of a cow (left) and a human (right). The white spot is the optic nerve and the black area is the most concentrated vision cones, with a protrusion of vision cells on the cow's retina

humans are nearly blind. Still, it is almost impossible to sneak up on a cow if the animal has even the slightest view of the intruder.

Once I was trying to photograph a heifer giving birth on the Felton Ranch, and I found out how well cows can see. I would sit in the truck next to the heifer's paddock and wait until the heifer, who was clearly in labor, would lie down. As soon as I got out of the truck and approached the fence, she would get up and move to the farthest part of the paddock. Although being a heifer she had no idea what was happening, she was absolutely sure she didn't want to be near a human being. At one side of the paddock sat the barn and some attached pens fenced tightly with vertical boards instead of the usual post and rail of the classic corral. The third time she moved away, she lay down near the barn. It looked as if it would be easy to drive over to the barn, enter one of the boarded-in pens, and sneak up close enough to take a picture just by holding the camera above the fence, aiming it in the general direction of the heifer. Anyone watching this performance would get a very good idea of cow eyesight. Shortly after I went through the barn and made my way to the fenced pen, every heifer in the pasture was staring at the fence. They had seen movement from up to two hundred yards away through cracks no more than an eighth of an inch (3 millimeters) wide. And the cow in labor got up and moved again. It was obviously time to stop harassing her at this delicate moment, so I gave up. A few minutes later the calf started to come and the heifer paid no attention to anything but the work at hand.

Sources

THE muddled history of (or speculation about) the domestication of cattle is scattered through a voluminous and somewhat contentious periodical literature, which has been selectively melded into this text. Perhaps the most persuasive account of domestication makes up just a part of Jared Diamond's Pulitzer Prize–winning *Guns, Germs, & Steel: The Fates of Human Societies* (New York: W. W. Norton, 1997). An overview of the issues is available in *The Evolution of Domestic Animals* by Ian Mason (London: Longman's, 1984). A useful introduction to paleobiology is in Simon M. Davis, *The Archaeology of Animals* (New Haven: Yale University Press, 1987). The only printed account of the self-taming behavior of a wild quadruped, a likely step in the domestication of the cow, is found in Valerius Geist, *Mountain Sheep and Man in the Northern Wilds* (Ithaca: Cornell University Press, 1975).

Modern genetic research into the ancestry and differentiation of cattle species is still lacking a synthesized study, but Dan Bradley of Trinity College, Dublin, and his colleagues are exploring the frontiers of evolutionary cattle genetics. A good start would be to locate D. G. Bradley, R. T. Loftus, P. Cunningham, and D. E. MacHugh, "Genetics and domestic cattle origins," in *Evolutionary Anthropology* (1998): 79–86. The aurochs, because it survived into the early modern world, is included in the monumental *Animal Encyclopedia* begun by Bernhard Grzimeks and still in revision and publication, most recently by Simon & Schuster, New York. The *Encyclopedia* provides such delightful arcana as the dimensions and mass of all the members of the vertebrate animal kingdom from aardvark to zebra. The modern Spanish fighting bull is the subject of a vast literature, but two books, Angus MacNab's *Fighting Bulls: An Account of the Bullfight* (New York: Harcourt, Brace, 1959) and Barnaby Conrad's *The Encyclopedia of Bull Fighting* (Boston: Houghton-Mifflin, 1961), are excellent and complementary, with little overlapping information.

There are three charming old books describing the state of the British cattle industry just as the various breeds were being defined and improved, and in some cases, subsequently allowed to disappear altogether. *Cattle: Breeds, Management and Diseases* by W[illiam] Youatt (London: A. O. Moore, 1858) is a marvelously opinionated work by one of England's most famous and prolix veterinary surgeons. His books on dogs, horses, and even pigs are more available than his work with cattle. A look at the state of cattle just before the improvements of the nineteenth century is found in *Observations on Live Stock* by George Culley (London, 1786). A similar survey, both written and illustrated by T[homas] Bewick, is *A General History of Quadrupeds* (Newcastle upon Tyne, 1791). Just as the many variations of so-called wild Park cattle were disappearing, the Reverend John Storer, a breeder and judge of Shorthorn cattle, produced *The Wild White Cattle of Great Britain, an Account of Their Origins, History and*

Present Status (London, Paris, and New York: Cassell Petter & Galpin, 1879). Storer is vastly credulous but equally entertaining (especially for those bemused by cattle in general). An argumentative book, rather disdainful of both Scots and Scottish cattle, written at the very height of English self-satisfaction (just prior to World War I), is Richard Lydekker's *The Ox and Its Kindred* (London: Methuen & Co., Ltd., 1912).

The early history of the Aberdeen-Angus breed is comprehensively dealt with in *The History of Polled Aberdeen or Angus Cattle* by James Macdonald and James Sinclair (Edinburgh: William Blackwood & Sons, 1882). Macdonald and Sinclair also provide a more thorough and accurate accounting than any other authors of the first exports of Aberdeen-Angus to North America. A more available history is the 1958 publication commissioned by the Aberdeen-Angus Society of Great Britain, *The Aberdeen-Angus Breed, a History* by James R. Barclay and Alexander Keith (Aberdeen, 1958). The earlier book by Macdonald and Sinclair is worth seeking out. With a shorter time span to cover, it is much richer in anecdotes of the early days. Another source of arcane information about the breed is the *Aberdeen-Angus Review*, published by the Aberdeen-Angus Society of Great Britain. It was and still is the model of what a society's publication should be, not entirely confined to the business at hand but willing to publish the ephemeral stories that inform and amuse the reader. A little-known and rare book (and the only autobiography by one of the founders of the breed) is William M'Combie's *Cattle and Cattle Breeders* (Edinburgh and London: Blackwood, 1886). Although some of his practices would not apply in the modern world, much of his advice on breeding and raising cattle is as wise as it was the day it was written. I am extremely indebted to a history of England in the nineteenth century that uses animal husbandry as a key to understanding that peculiar society—Harriet Ritvo's *The Animal Estate: The English and other Creatures in the Victorian Age* (Cambridge: Harvard University Press, 1987). It is rich in detail on the more bizarre aspects of the cow's world, especially the obscenely overweight animals entered in prize competitions.

If the introduction of Aberdeen-Angus to the United States proceeded by fits and starts, the same can be said for Canada. The official history *Aberdeen-Angus Cattle in Canada* by a Mr. F. W. Crawford tells the Canadian version. Perhaps characteristic of that nation, it is a rather quiet or subdued argument for primacy, acknowledging up front that their first introduction had as little effect on the course of history as George Grant's importations to the United States.

.   .   .

For an account of the origins of New World cattle ranching, I have relied greatly on a truly wide-ranging work: Terry G. Jordan's *North American Cattle-Ranching Frontiers: Origins, Diffusion and Differentiation* (Albuquerque: University of New Mexico Press, 1993). The specifically Celtic practices he describes are the foundations for the centuries of cattle-trading between the Highlands and England proper, an industry described in charming detail by A.R.B. Haldane in *The Drove Roads of Scotland* (London: Thomas Nelson and Sons, 1952). There are several reprints available. For the final transformation from the near-mythical and brief era of the open range in North America, several sources are enlightening. Any discussion of the western cattle range begins with Walter Prescott Webb's *The Great Plains* (Boston: Ginn & Co., 1931, or a facsimile of that edition from the University of Nebraska Press, Lincoln, 1981). Webb's geographical determinism is currently out of fashion, but his facts are not in dispute. A dispassionate discussion of the closing of the open range is found in Edward Everett Dale's *The Range Cat-*

tle Industry: Ranching on the Open Plains from 1865 to 1925 (Norman: University of Oklahoma Press, 1930, reprinted there in 1960). The great closure in the immediate vicinity of the Felton Angus Ranch is detailed in Robert Fletcher's article "End of the open range in Eastern Montana," *Mississippi Valley Historical Review* 16 (1929–1930). Many authors describe the boom mentality of the early open-range ranchers, but Walter, Baron von Richthofen (uncle of the World War I German ace, the Red Baron, Manfred von Richthofen), was perhaps the most egregious enthusiast at describing the inevitable incredible wealth that ranchers could expect in his *Cattle-Raising on the Plains of North America,* self-published in 1885, just as the first of the great blizzard winters destroyed the open-range herds. The University of Oklahoma Press reprinted it in 1964 and 1969. The Great Sioux War that opened the range was fought and won out of Miles City, Montana, which was named for the Army colonel who was the single successful campaigner in the entire war. This is recounted in *Yellowstone Command: Colonel Nelson A. Miles and the Great Sioux War 1876–1877* (Lincoln and London: University of Nebraska Press, 1991). A larger view of the man, including his later career, is in *Nelson A. Miles and the Twilight of the Frontier Army* by Robert Wooster (University of Nebraska Press, 1993).

The unique nature of a cow's vision (and that of several other vertebrates) is summarized in "The Topography of Vision in Mammals of Contrasting Life Style: Comparative Optics and Retinal Organization" by A. Hughes, in *The Visual System in Vertebrates,* Frederick Crescitelli, ed. (New York: Springer-Verlag, 1977).

Although cattle can be raised for market in any number of systems, a good overview of what beef cows actually need to eat is in Tilden Wayne Perry's *Beef Cattle Feeding and Nutrition* (New York: Academic Press, 1980). Perry's emphasis is on feedlot management, which accounts for almost all of North American beef production. A look at the British system, typical of Aberdeen-Angus husbandry in their native country, is included in *Bovine Medicine,* A. H. Andrews, ed. (London: Blackwell, 1992).

The ethology of cattle is much less explored than the genetically enabled behavior of conveniently small things like prairie voles (valued because they are a uniquely monogamous rodent) or large and exotic creatures like elephants and gorillas. What little research on cattle behavior there is tends to concentrate on confined cattle, that is, dairy breeds, Holstein in particular, which are rather lethargic beasts. Fortunately, all mammals appear to share common traits in their neurobiology so that clinical studies of nursing human mothers and lactating laboratory rats find similar responses to internal hormones and external stimulations. The roles of oxytocin in behavior and physiology are thoroughly and repeatedly studied but less commonly translated into lay language. A good place to begin is an article, and the accompanying bibliography, by Thomas R. Insel, M.D.: "A Neurobiological Basis of Social Attachment" in the *American Journal of Psychiatry* 154 (June 1997). There is considerable discussion of oxytocin and the other behavior-affecting hormones in *The Integrative Neurobiology of Affiliation,* C. Sue Carter et al., eds., in *Annals of the New York Academy of Sciences,* vol. 807, published by the academy in New York, 1997.

One of the most discernible aspects of mammalian behavior is the enormous role that odor plays in mating and maternal behavior. Cattle (and a host of other creatures, including snakes) react specifically to pheromones, the veritable cupid's arrow of the nose. An entertaining and lucid description can be found in Lyall Watson's *Jacobson's Organ and the Remarkable Nature of Smell* (New York and London: W. W. Norton & Company, 1999).

Index